ENVIRONMENTAL HEALTH -
PHYSICAL, CHEMICAL AND BIOLOGICAL FACTORS

OUTDOOR RECREATION IN THE NORTHERN UNITED STATES AND PROJECTED OUTLOOK TO 2060

Environmental Health - Physical, Chemical and Biological Factors

Additional books in this series can be found on Nova's website under the Series tab.

Additional E-books in this series can be found on Nova's website under the E-book tab.

Environmental Health -
Physical, Chemical and Biological Factors

Outdoor Recreation in the Northern United States and Projected Outlook to 2060

Alexander N. Borden
Editor

Copyright © 2014 by Nova Science Publishers, Inc.

All rights reserved. No part of this book may be reproduced, stored in a retrieval system or transmitted in any form or by any means: electronic, electrostatic, magnetic, tape, mechanical photocopying, recording or otherwise without the written permission of the Publisher.

For permission to use material from this book please contact us:
Telephone 631-231-7269; Fax 631-231-8175
Web Site: http://www.novapublishers.com

NOTICE TO THE READER

The Publisher has taken reasonable care in the preparation of this book, but makes no expressed or implied warranty of any kind and assumes no responsibility for any errors or omissions. No liability is assumed for incidental or consequential damages in connection with or arising out of information contained in this book. The Publisher shall not be liable for any special, consequential, or exemplary damages resulting, in whole or in part, from the readers' use of, or reliance upon, this material. Any parts of this book based on government reports are so indicated and copyright is claimed for those parts to the extent applicable to compilations of such works.

Independent verification should be sought for any data, advice or recommendations contained in this book. In addition, no responsibility is assumed by the publisher for any injury and/or damage to persons or property arising from any methods, products, instructions, ideas or otherwise contained in this publication.

This publication is designed to provide accurate and authoritative information with regard to the subject matter covered herein. It is sold with the clear understanding that the Publisher is not engaged in rendering legal or any other professional services. If legal or any other expert assistance is required, the services of a competent person should be sought. FROM A DECLARATION OF PARTICIPANTS JOINTLY ADOPTED BY A COMMITTEE OF THE AMERICAN BAR ASSOCIATION AND A COMMITTEE OF PUBLISHERS.

Additional color graphics may be available in the e-book version of this book.

Library of Congress Cataloging-in-Publication Data

ISBN: 978-1-63117-110-9

Published by Nova Science Publishers, Inc. † New York

CONTENTS

Preface   **vii**

Chapter 1   Outdoor Recreation in the Northern United States   **1**
                 H. Ken Cordell, Carter J. Betz, Shela H. Mou
                 and Dale D. Gormanson

Chapter 2   Outlook for Outdoor Recreation in the Northern
                 United States. A Technical Document Supporting the
                 Northern Forest Futures Project with
                 Projections through 2060   **67**
                 J. M. Bowker and Ashley E. Askew

Index   **127**

PREFACE

Outdoor recreation can take many forms depending on the types of activities, settings, social engagements, equipment, and times chosen by the recreation participant. Recreation can be physically active (for example, hiking) or more sedentary (for example, viewing natural scenery). Many of the activities of interest to the Renewable Resources Planning Act (RPA) Assessment and to assessments of current or future northern forest conditions are classified as "nature-based" in that they are in some way associated with wildlife, birds, streams, lakes, snow and ice areas, trails, rugged terrain, mountains, caves, and other natural outdoor resources or settings. This book discusses outdoor recreation in the norther United States and provides a technical document supporting the northern forest futures project that includes projections through 2060.

Chapter 1 - In the last two decades, the North's population grew at a considerably slower rate than the Nation as a whole. Nevertheless, this region's population is large and in all likelihood will continue to grow. This means greater development of land and water resources at the same time that there is growth in demand for outdoor recreation. This report looks at recent population trends and forecasts within the context of other U.S. regions, demographic composition of population, recreation participation by residents age 16 and older, trends in activities and time spent outdoors by its youth, and the changes occurring in recreation resources, both public and private. The region referenced here includes the area within the corner States of Maine, Minnesota, Missouri, and Maryland. Much of the research reported here ties to data, analyses, and findings developed for the U.S. Department of Agriculture Forest Service 2010 Renewable Resources Planning Act (RPA) Assessment (Cordell 2012) and how they affect the sustainability of northern forests.

Chapter 2 – The authors developed projections of participation and use for 17 nature-based outdoor recreation activities through 2060 for the Northern United States. Similar to the 2010 Resources Planning Act (RPA) assessment, this report develops recreation projections under futures wherein population growth, socioeconomic conditions, land use changes, and climate are allowed to change over time.

Findings indicate that outdoor recreation will likely remain a key part of the region's future social and economic fabric. The number of participants in 14 of the 17 recreation activities is projected to increase over the next five decades. In about two-thirds of 17 activities, the participation rate will likely decrease, but population growth would ensure increases in the number of adult participants. Some climate futures could lead to participant decreases for certain activities. Hunting, snowmobiling, and undeveloped skiing appear to be the only activities for which a decrease in participants is likely. Total days of participation would generally follow the pattern of participant numbers. With the exceptions of hunting, visiting primitive areas, and whitewater activities, snowmobiling, undeveloped skiing, total days are expected to increase for the remaining 14 activities, some less so than others because of climate differences.

In: Outdoor Recreation ...
Editor: Alexander N. Borden

ISBN: 978-1-63117-110-9
© 2014 Nova Science Publishers, Inc.

Chapter 1

OUTDOOR RECREATION IN THE NORTHERN UNITED STATES[*]

H. Ken Cordell, Carter J. Betz, Shela H. Mou and Dale D. Gormanson

ABSTRACT

In the last two decades, the North's population grew at a considerably slower rate than the Nation as a whole. Nevertheless, this region's population is large and in all likelihood will continue to grow. This means greater development of land and water resources at the same time that there is growth in demand for outdoor recreation. This report looks at recent population trends and forecasts within the context of other U.S. regions, demographic composition of population, recreation participation by residents age 16 and older, trends in activities and time spent outdoors by its youth, and the changes occurring in recreation resources, both public and private. The region referenced here includes the area within the corner States of Maine, Minnesota, Missouri, and Maryland. Much of the research reported here ties to data, analyses, and findings developed for the U.S. Department of Agriculture Forest Service 2010 Renewable Resources Planning Act (RPA) Assessment (Cordell 2012) and how they affect the sustainability of northern forests.

---

[*] This is an edited, reformatted and augmented version of The United States Department of Agriculture, Forest Service, Northern Research Station publication, Gen. Tech. Rep. NRS-100, dated August 2012.

INTRODUCTION

This publication is part of the Northern Forest Futures Project, through which the Northern Research Station of the U.S. Forest Service examines the issues, trends, threats, and opportunities facing the forests of the northern United States. It complements the *Forests of the Northern United States* (Shifley et al. 2012) which summarizes forest-related concerns that are unique to the northern United States and discusses characteristics associated with forest sustainability in the region.

This publication is intended for natural resource managers and planners, policy makers, State natural resource agencies, politicians, students, and those who want to know more about recreation in northern forests.

In this report, we describe recent population trends and forecasts for the North within the context of other U.S. regions, demographic composition of its population, recreation participation by its residents age 16 and older, trends in activities and time spent outdoors by its youth, and the recreation resources, both public and private. The region referenced here includes the 20 states bounded by the corner states of Maine, Minnesota, Missouri, and Maryland.

Much of the research reported here ties to data, analyses, and findings developed for the U.S. Department of Agriculture Forest Service 2010 Renewable Resources Planning Act (RPA) Assessment (Cordell 2012). The data and methods employed are described in the appendix at the end of this report. The Forest and Rangeland Renewable Resources Planning Act of 1974 mandated a decennial national assessment (with periodic updates) of the renewable resources on all public and private forest and range ownerships. Each RPA Assessment provides a snapshot of current conditions and trends on U.S. forest and range lands, identifies factors that drive change, and makes model-driven 50-year projections of demands, uses, and conditions for recreation, water, timber, wildlife (biodiversity), and urban-forest and range resources. Trends and forecasts in land use and climate change are also included. The 2010 RPA Assessment stresses the influence of climate change on forest and grassland resources and has adapted three socioeconomic scenarios based on the fourth world assessment of climate change (Intergovernmental Panel on Climate Change 2007). RPA population forecasts to 2060 based on these three scenarios are reported later in this report. The 2010 set of special RPA resource studies (which include 2060 forecasts) and the national summary are in press.

Historical Context

It was not until the post-World War II years that a number of social and economic forces combined to make outdoor recreation a national phenomenon that required serious attention and study. Three of the major forces at work included rising real incomes, the proliferation of automobiles and highways (especially the Interstate Highway System), and increasing leisure as the United States continued shifting from a predominantly agricultural to a manufacturing and service-based economy. Increasingly, in the 1950s and 1960s, Americans took to the open road to see and experience "the great outdoors." A direct result was mounting pressures on recreation facilities and most public lands (Clawson and Knetsch 1966, Cordell 2012). Consequently, major efforts were undertaken beginning in the late 1950s to better understand Americans' growing interest in outdoor recreation. Most notable was the Outdoor Recreation Resources Review Commission, established by Congress in 1958 to conduct a comprehensive nationwide assessment of outdoor recreation conditions and trends.

Interest in monitoring outdoor recreation trends continues to the present day (Cordell 2008). In an earlier national report, we reported that Americans' participation in outdoor activities, including nature-based recreation activities, had been increasing up through the first few years of the 2000s (Cordell et al. 2004).

Overall, since the commission released its report (Outdoor Recreation Resources Review Commission 1962), many forms of outdoor activity and public land visitation have been observed to be growing and diversifying:

Both the NSRE (National Survey on Recreation and the Environment) and the National Survey on Fishing, Hunting, and Wildlife-Associated Recreation show that participation in some nature-based activities has declined. However, for many other activities there seems to be growing popularity. Some outdoor recreation activities have even demonstrated rather strong popularity growth. One such activity is visiting wilderness and other primitive areas. (Cordell et al. 2008)

Because trends in outdoor recreation have far reaching implications for both people and natural resources, a close look at those trends and projected futures for the Northern States is an important part of the Northern Forest Futures Project, currently underway at the Forest Service (Northern Research Station, Eastern Region, Forest Products Laboratory, and Northeastern Area State and Private Forestry) in partnership with the Northeastern Area Association of State Foresters and the University of Missouri.

Outdoor Recreation Defined

Outdoor recreation can take many forms depending on the types of activities, settings, social engagements, equipment, and times chosen by the recreation participant. Recreation can be physically active (for example, hiking) or more sedentary (for example, viewing natural scenery). Many of the activities of interest to the RPA Assessment and to assessments of current or future northern forest conditions are classified as "nature-based" in that they are in some way associated with wildlife, birds, streams, lakes, snow and ice areas, trails, rugged terrain, mountains, caves, and other natural outdoor resources or settings. For example, included among our list of nature-based activities are mountain biking, coldwater fishing, whitewater rafting, downhill skiing, primitive camping, backpacking, mountain climbing, visiting prehistoric sites, saltwater fishing, and snorkeling. Nature-based recreation participation is summarized for the North across seven activity groups:

- **Visiting recreation and historic sites**— Visiting the beach, visiting prehistoric sites, visiting historic sites, developed camping, swimming in lakes/ponds/streams, and visiting watersides (besides beaches)
- **Viewing/photographing nature**—Viewing/ photographing birds, fish, other wildlife, natural scenery, wildflowers/trees/other plants, visiting nature centers, sightseeing, gathering mushrooms/berries, and participating in boat tours or excursions
- **Backcountry activities**—Backpacking, day hiking, horseback riding on trails, mountain climbing, visiting a wilderness or primitive area, primitive camping, mountain biking, caving, rock climbing, and orienteering
- **Motorized activities**—Motorboating, offhighway-vehicle driving (four-wheel-drive vehicle, all-terrain vehicle, or motorcycle), snowmobiling, using personal watercraft, and waterskiing
- **Hunting and fishing**—Anadromous fishing (salt-to-fresh-water migratory fish such as salmon), coldwater fishing, warmwater fishing, saltwater fishing, big game hunting, small game hunting, and migratory bird hunting
- **Non-motorized boating and diving**— Canoeing, kayaking, rafting, rowing, sailing, surfing, windsurfing, snorkeling, and scuba diving
- **Snow skiing and other winter activities**— Cross country skiing, downhill skiing, snowboarding, snowshoeing, and ice fishing

POPULATION AND DEMOGRAPHIC TRENDS

Current Population Trends for the North

Race and ethnic composition data from the U.S. Department of Commerce, Bureau of the Census, along with the percentage change trends from 1990 to 2009, are summarized by region in Table 1. Race and ethnicity are important determinants of what people choose as outdoor recreation activities and the settings they prefer for those activities. For example, African Americans tend to participate much less frequently in wildland recreation activities and many Hispanics appear to prefer settings that will accommodate large-group activities such as picnicking (Cordell et al. 2004).

The changes in the racial and ethnic makeup of the U.S. population have been dramatic since the 1990 census. Although all groups have been growing in number, generally, Hispanics and Asian/Pacific Islanders have been growing fastest. Slowest growing of all the groups has been non-Hispanic Whites. The North had the lowest growth rate and the Rocky Mountains highest. The highest percentage growth of any group since 1990 has been Asians/Pacific Islanders in the Rocky Mountains and South, each with about a 187 percent increase. Non-Hispanic Whites experienced slight population losses in the North and Pacific Coast.

The Rocky Mountains and South are the only regions that exceeded the national rate for all groups. Total population growth in the North was 10.7 percent, less than half the national rate (23.4 percent); this held true for all groups except Asians/Pacific Islanders, which more than doubled in the North since 1990 (122.7 percent). The non-Hispanic White population almost held constant, declining just 0.2 percent. However, the North has the largest share of non-Hispanic Whites (almost 74 percent), which depressed its overall growth rate to less than 11 percent. About 14.9 million African Americans live in the North, 40 percent of the national total; the growth rate for the group was 19.7 percent, compared to 28.6 percent nationally.

The North lags behind other regions in American Indian population, although it is a close second to the Pacific Coast and has grown 10 percent faster than that region since 1990.

Only the Pacific Coast has more Asians/Pacific Islanders, whose growth rate was nearly twice as high in the North as the Pacific Coast. The Hispanic population in the North almost exactly doubled, growing to almost 11.1 million.

Table 1. Population in 2009 by race/ethnicity and region, and change since 1990

Race / ethnicity Non-Hispanic	North Population (thousands)	Change (percent)	South Population (thousands)	Change (percent)	Rocky Mountains Population (thousands)	Change (percent)	Pacific Coast Population (thousands)	Change (percent)	United States Population (thousands)	Change (percent)
White	92,333.8	-0.2	63,761.3	14.6	19,544.5	25.7	24,211.7	-1.7	199,851.2	6.1
African American	14,899.9	19.7	19,202.6	37.9	998.6	77.6	2,580.5	9.3	37,681.5	28.6
American Indian	421.7	24.7	716.4	38.8	779.5	40.2	443.2	14.7	2,360.8	31.4
Asian or Pacific Islander	4,806.0	122.7	2,626.0	186.4	732.2	187.4	5,970.4	62.9	14,134.6	102.0
Two or more races[a]	1,524.1	.	1,311.0	.	442.1	.	1,281.8	.	4,559.0	.
Hispanic[b]	11,064.6	100.1	16,696.6	153.6	5,700.5	167.3	14,957.7	84.0	48,419.3	116.4
Total	125,050.0	10.7	104,313.8	34.4	28,197.5	48.0	49,445.2	26.2	307,006.6	23.4

[a]Percentage change for two or more races is missing because U.S. citizens were not offered the option to select more than one race until the 2000 census.

[b]Hispanics of all races are included in this category.

Source: U.S. Department of Commerce, Bureau of the Census 1990, 2009a.

Table 2. Population in 2009 by age group and region, and change since 1990

Age Group (years)	North Population (thousands)	Change (percent)	South Population (thousands)	Change (percent)	Rocky Mountains Population (thousands)	Change (percent)	Pacific Coast Population (thousands)	Change (percent)	United States Population (thousands)	Change (percent)
<6	9,569.8	-2.3	9,022.6	29.9	2,603.7	40.2	4,289.1	12.9	25,485.2	13.8
6 to 10	7,886.2	0.0	7,151.7	25.4	2,009.3	28.4	3,323.3	14.9	20,370.5	12.9
11 to 15	8,088.2	9.3	6,857.8	27.5	1,893.1	34.2	3,321.7	29.8	20,160.9	20.4
16 to 24	15,725.2	4.2	13,165.7	23.2	3,675.3	47.1	6,377.4	17.4	38,943.7	15.5
25 to 34	16,138.0	-16.6	14,275.9	7.4	3,995.4	23.6	7,157.0	-2.1	41,566.3	-3.7
35 to 44	16,880.1	-0.5	14,126.6	23.2	3,635.1	27.2	6,888.2	11.7	41,530.0	10.9
45 to 54	19,028.4	64.1	14,688.1	88.1	3,856.5	111.0	7,019.5	83.3	44,592.5	77.9
55 to 64	14,740.7	47.0	11,597.0	75.8	3,092.5	102.9	5,356.8	80.5	34,786.9	64.7
≥65	16,993.4	14.4	13,428.4	38.0	3,436.6	51.1	5,712.1	35.3	39,570.6	27.3
Total	125,050.0	10.7	104,313.8	34.4	28,197.5	48.0	49,445.2	26.2	307,006.6	23.4

Source: U.S. Department of Commerce, Bureau of the Census 1990, 2009a.

Age Distribution

Age is another important determinant of recreation activity choices (Cordell et al. 2004). Similar to race and ethnicity, the age distribution of the U.S. population has been changing over time (Table 2). The fastest growing age group since 1990 (in percentage change) has been age 45 to 54, followed by age 55 to 64. Third fastest has been age 65 and older. The 45-to-54-age group grew fastest in all regions. In the North, the 25-to34-age group decreased nearly 17 percent, contributing to a national drop in population of almost 4 percent. The Pacific Coast was the only other region to lose population in this young adult segment. Also losing population in the North was the youngest age group (younger than 6 years). The 6-to-10-age group essentially held constant since 1990. These two youngest segments of the U.S. population grew at double-digit rates in every other region, fastest in the Rocky Mountains.

Similar to the Nation, northern Baby Boomers (ages 45 to 54 and 55 to 64) dominated all other age groups in percentage growth (Table 2), but at a slower rate. However, the North experienced a greater disparity between the two Baby Boomer groups and the third-place age group (age 65 and older) than the Nation as a whole. Percentage growth for the age 55 to 64 group was more than three times that of the oldest age group, and the age 45 to 54 group grew at more than four times the rate of the age 65 and older group. No other region nor the Nation approached these growth rates.

Further, the North has the oldest population of any U.S. region. Almost 41 percent of its residents are age 45 or older; no other region has more than 38 percent in this age group. As with race and ethnicity, the South and Rocky Mountains were the only regions to outpace the national growth rate for every single age group. Conversely, northern populations increased (or decreased) at a slower rate than the Nation in all age groups. The three age groups in the North that lost population include the youngest age group (under 6) and the two groups that spanned age 25 to 44. The number of young adults age 25 to 34 in their prime childbearing years, in particular, decreased at a far greater rate in the North than any other age group in any region. This trend helps explain the decrease and lack of growth in the two youngest age groups. In addition to growing more slowly than the Nation and any other region in total population, the North's modest 11 percent population gain since 1990 occurred overwhelmingly in the three oldest cohorts age 45 and older. A greater share of the northern population appears to be aging-inplace compared to other regions. The decrease in young adult populations in the North is the result of lower birth-to-death rates and of young people seeking economic opportunities elsewhere (Franklin 2003, Yang and Snyder 2007).

Population Density

Population density (persons per square mile) is greatest in Florida, in the Piedmont areas of North Carolina to Georgia, along the coast of the northern Atlantic States, in several Great Lakes and midwestern metropolitan areas, in eastern Texas, in the Denver-Front Range area, and in scattered areas along the Pacific Coast and into Arizona (Fig. 1). In Alaska, density is greatest in the Anchorage area.

The North has for years been well known for the cluster of densely populated counties that extend from the Washington-Baltimore metropolitan to southern New Hampshire. Urban or mostly urban counties also stretch almost continuously along the Great Lakes from central New York to Green Bay, WI. Other high-density areas include western Pennsylvania, parts of Ohio, and metropolitan areas around Indianapolis, St. Louis, the Twin Cities of Minneapolis/St. Paul, and Kansas City, MO.

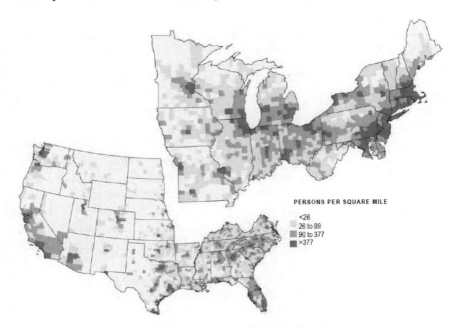

Figure 1. Population density by county in the contiguous United States, 2009 (Source: U.S. Department of Commerce, Bureau of the Census 2009b).

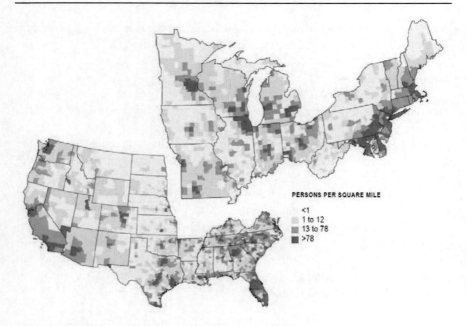

Figure 2. Change in population density by county in the contiguous United States, 1990 to 2009 (Source: U.S. Department of Commerce, Bureau of the Census 1990).

The North has the largest number of counties in the two most densely populated classes (more than 90 persons per square mile); most of its counties in the least densely populated category are located in the midwestern area and northern Maine (Fig. 1).

Figure 2 shows that much of the overall population-density growth in the East has occurred along the northern Atlantic coast, down the Piedmont and Southern Appalachians from North Carolina to greater Atlanta, in peninsular Florida, around Chicago, the Twin Cities, and the major cities of Texas.

Elsewhere in the United States, growth occurred in the Denver and Salt Lake City metropolitan areas, in the Bay Area and southern California areas, and in greater Seattle and Portland, OR. In some areas— such as eastern Texas, metropolitan Atlanta, and Orange County in California—population growth exceeds 500 persons per square mile, which is the U.S. Department of Commerce, Bureau of the Census definition of an urban area. Greater concentrations of people in places near public lands and bodies of water are likely to put increasing pressures on these limited resources.

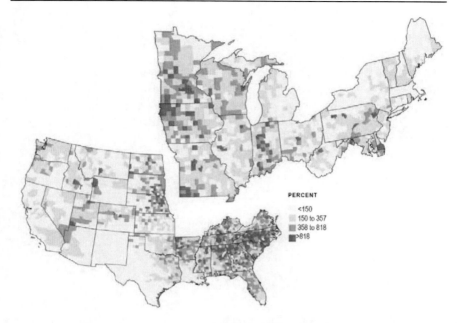

Figure 3. Change in Hispanic population of all races by county in the contiguous United States, 1990 to 2009 (source: U.S. Department of Commerce, Bureau of the Census 1990, 2009b).

In the North, population density increased the most throughout the Washington-to-Boston urban corridor and also in the greater Chicago and Twin Cities areas. Noticeably different from their high population-density rankings (Fig. 1), few Northern counties are in the two highest growth categories (which represents the top 30 percent of all U.S. counties); this is especially true in Ohio, Michigan, and Indiana. Conversely, more Northern counties (including much of Pennsylvania, West Virginia, New York, Illinois, and Iowa) are in the lowest growth category, which represents a loss or negligible growth in population since 1990. The urban cores of metropolitan Detroit, Cleveland, Cincinnati, St. Louis, and Buffalo, NY, lost population over the 19-year period.

Hispanics

From 1990 to 2009, Hispanic population growth has exceeded 800 percent in some U.S. counties (Fig. 3). Much of the fastest growth has been in the southern States bordering the Atlantic Ocean and Mississippi River. High rates of growth have also occurred through the upper midwestern area and through selected areas of the West. With some exceptions, the rate of Hispanic growth in the North has been lower than most of the South; however, even the second-lowest category represents up to 357 percent growth.

Just a few northern counties—in Minnesota, Missouri, West Virginia, and Illinois—have experienced reductions in their Hispanic populations. Very high rates of Hispanic population growth—more than 818 percent in less than 20 years—occurred in Minnesota, Iowa, Missouri, Indiana, and a scattering of counties elsewhere. The lowest rates occurred in Michigan, New York, much of Ohio and Illinois, and the New England States.

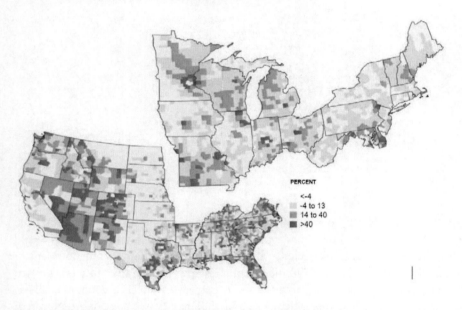

Figure 4. Change in non-Hispanic White population by county in the contiguous United States, 1990 to 2009 (source: U.S. Department of Commerce, Bureau of the Census 1990, 2009b).

Non-Hispanic Whites

The non-Hispanic White population in the United States has been growing in metropolitan areas such as Atlanta, Washington, the Raleigh/Durham area in North Carolina, the Twin Cities, eastern Texas, and throughout much of the West (Fig. 4).

Areas rich in natural amenities—such as the Rocky Mountains, Florida, Arizona, Colorado, Utah, and Nevada—appear to have the fastest growth of non-Hispanic White populations. The top tier of percentage change includes counties that increased more than 40 percent, much lower than the highest level of Hispanic population growth. In the North, these high-growth counties were relatively few, located mostly in suburban areas around major cities. Three other areas with faster growing non-Hispanic White populations that are not highly urbanized, but also possess abundant natural amenities, are the Delmarva Peninsula region in Delaware and Maryland, the Delaware Water Gap area of northeastern Pennsylvania, and southern Missouri.

Population Projections for Three Growth Scenarios (2008 to 2060)

Similar to the trends in population growth and composition since 1990, the regions likely to lead the Nation in projected rate of change under the moderate growth scenario are the Rocky Mountains at 76 percent and the South at 57 percent (Table 3). The Pacific Coast follows closely at 54 percent. The North lags behind the others by a wide margin, with just 26 percent expected growth. The intermountain area of the Rocky Mountains far exceeds all other areas with projected growth of 89 percent (nearly three times the rate of the Great Plains area). By 2060, the South is expected to pass the North as the Nation's most populous region. Currently (2009), the North accounts for 40.7 percent of the total U.S. population, but is projected to drop to 35.2 percent of the total by 2060, compared to 36.6 percent (up from 34.0 percent) in the South, 11.1 percent (up from 9.2 percent) in the Rocky Mountains, and 17.1 percent (up from 16.1 percent) in the Pacific Coast.

The eight States in the north-central area (west of and including Ohio) are projected to grow just slightly faster (26.8 percent) than both the region as a whole (26.0 percent) and the 12 States and the District of Columbia that comprise the northeastern area (25.3 percent). The northeastern area, however, has 7 of the top 10 States ranked by percentage growth, led by New Hampshire, Maryland, and Vermont, each with more than 50 percent projected growth. Minnesota is the only State in the north central area expected to grow more than 50 percent. Ohio, West Virginia, and New York are the three lowest ranking States, each expected to grow less than 13 percent. The District of Columbia is projected to lose almost 17 percent of its population.

Table 3. Estimated population for 2009, projections to 2060 by region and State for three population growth scenarios, and percentage change from 2009 to 2060 for the moderate growth projection

Region, State	Population 2009	Moderate growth[a] 2060	Change from 2009	High growth[b] 2060	Low growth[c] 2060
	(thousands)	(thousands)	(percent)	(thousands)	(thousands)
Northern States	125,050.0	157,597.9	26.0	178,045.6	139,964.2
Connecticut	3,518.3	4,280.8	21.7	4,836.2	3,801.8
Delaware	885.1	1,308.6	47.8	1,478.4	1,162.2
District of Columbia	599.7	499.7	-16.7	564.5	443.8
Illinois	12,910.4	16,364.5	26.8	18,487.7	14,533.5
Indiana	6,423.1	8,147.5	26.8	9,204.6	7,235.9
Iowa	3,007.9	3,612.9	20.1	4,081.7	3,208.7
Maine	1,318.3	1,755.5	33.2	1,983.2	1,559.0
Maryland	5,699.5	9,120.0	60.0	10,303.3	8,099.5
Massachusetts	6,593.6	7,801.1	18.3	8,813.2	6,928.2
Michigan	9,969.7	12,173.3	22.1	13,752.7	10,811.2
Minnesota	5,266.2	7,987.7	51.7	9,024.1	7,094.0
Missouri	5,987.6	8,091.3	35.1	9,141.1	7,186.0
New Hampshire	1,324.6	2,255.5	70.3	2,548.1	2,003.1
New Jersey	8,707.7	11,969.2	37.5	13,522.2	10,630.0
New York	19,541.5	21,929.1	12.2	24,774.3	19,475.4
Ohio	11,542.6	12,811.1	11.0	14,473.3	11,377.7
Pennsylvania	12,604.8	15,235.9	20.9	17,212.7	13,531.2
Rhode Island	1,053.2	1,403.6	33.3	1,585.7	1,246.6
Vermont	621.8	956.2	53.8	1,080.3	849.3
West Virginia	1,819.8	2,033.8	11.8	2,297.6	1,806.2
Wisconsin	5,654.8	7,860.6	39.0	8,880.5	6,981.1
Southern States	104,313.8	163,673.8	56.9	184,909.9	145,360.3
Rocky Mountains States	28,197.5	49,695.6	76.2	56,143.5	44,135.2
Pacific Coast States	49,445.2	76,340.6	54.4	86,245.5	67,798.9
U.S. total	307,006.6	447,308.0	45.7	505,344.5	397,258.6

a – c: The moderate growth scenario corresponds to mid-range population growth to about 447 million people by 2060, and an average personal income of around $73,000. The high growth scenario projects the highest population growth, reaching more than 505 million and the lowest projected average personal around $50,000. The low growth scenario projects the lowest population growth and mid-level personal income, predicting a population of 397 million people with average personal income around $54,000 by 2060.

Source: Cordell 2012, U.S. Department of Commerce, Bureau of the Census 2009a.

Figures 5 through 7 show the geographic patterns of projected changes in population density by 2060—ranging from lowest (fewer than 2 persons per square mile) to the highest (more than 186 persons per square mile)— for the low (Fig. 5), moderate (Fig. 6), and high (Fig. 7) population growth scenarios. For the purposes of this analysis, land area in all counties is assumed to remain constant.

Figure 5. Change in persons per square mile by county in the contiguous United States, 2009 to 2060, for a low growth population projection (Sources: Cordell 2012, U.S. Department of Commerce, Bureau of the Census 2009b). (The low growth scenario projects the lowest population growth and mid-level personal income, predicting a population of 397 million people with average personal income around $54,000 by 2060).

Immediately apparent in the low-growth projection scenario (Fig. 5) is the presence of numerous lower density counties distributed throughout the North, especially throughout much of the midwestern area, New York State, and upper New England. The highest-growth counties, which are expected to add more than 186 persons per square mile, are concentrated mainly in the Washington-to-Boston urban corridor and in other suburban areas throughout the region.

The second-tier counties are mostly located around those counties with highest growth, as well as in Michigan, Wisconsin, or southern Missouri.

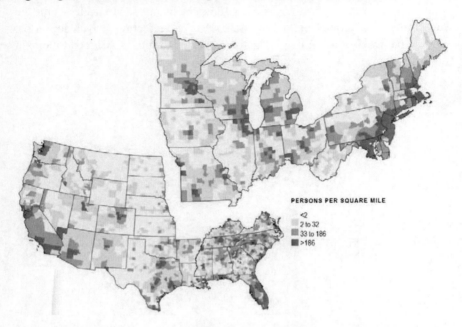

Figure 6. Change in persons per square mile by county in the contiguous United States, 2009 to 2060, for a moderate growth population projection (Source: Cordell 2012, U.S. Department of Commerce, Bureau of the Census 2009b). (The moderate growth scenario corresponds to mid-range population growth to about 447 million people by 2060, and an average personal income of around $73,000).

The moderate growth projection scenario (Fig. 6), which closely approximates the Census Bureau State projections, has fewer low-growth counties, as expected, and more counties in the intermediate ranges (2 to 186 additional persons per square mile). The highest-growth projection scenario mostly adds to the clusters of counties around the major urban centers, especially those in the northeastern area and those near Chicago, Detroit, and the Twin Cities. Under this scenario, very few high-growth counties in the North were added to nonmetropolitan areas high in natural amenities.

Under the high-growth scenario (Fig. 7), more counties shift from the lowest to the two moderate growth categories. The highest growth counties that are expected to add significant population density of more than 186 persons per square mile appear to be limited almost entirely to metropolitan areas, with only a few exceptions. A number of counties scattered throughout the region are projected to remain in the lowest growth class (fewer than 2 persons per square mile added, including population losses), especially in Iowa, northern Missouri and eastward to West Virginia, plus several counties northeastward from Pennsylvania to Maine.

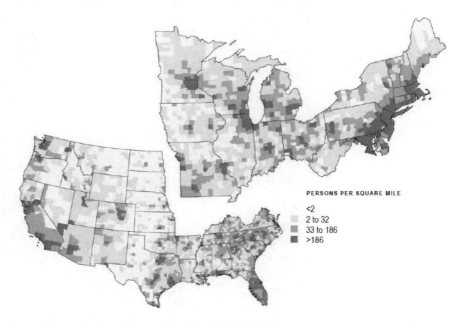

Figure 7. Change in persons per square mile by county in the contiguous united states, 2009 to 2060, for a high growth population projection (Source: Cordell 2012, U.S. Department of Commerce, Bureau of the Census, 2009b). (The high growth scenario projects the highest population growth, reaching more than 505 million and the lowest projected average personal around $50,000).

OUTDOOR RECREATION TRENDS

From 1999 to 2009 (single year labels which represent pooled data from the two data collection periods of 1999 to 2001 and 2005 to 2009), the number of people age 16 and older who participated in outdoor recreation grew by 8.5 percent nationally, from an estimated 208.2 million to 226.0 million (Fig. 8).

A participant is anyone who engaged in one or more of 60 outdoor activities during the past 12 months. Included in the list of 60 was a wide range of activities such as attending family gatherings outdoors, viewing wildlife and birds, backpacking, and mountain climbing. Across the range of these activities, the indexed number of total annual activity days of participation (measured as the product of the average number of days per activity times the number of participants and then summed across all activities) increased 32.8 percent from 61.8 billion to 82.0 billion. Average annual days of participation per person increased about 22 percent, from roughly 297 to about 363 total activity days per person per year. (These numbers may seem high, but they represent participation in more than one activity during any given day. So, these averages for "activity days" are sum totals across activities.)

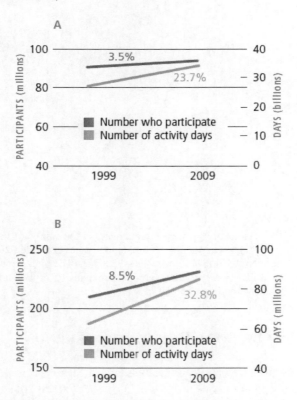

Figure 8. Growth in the number of participants and the number of participation days in 60 outdoor recreation activities in (A) the north and (B) the united states, 1999 to 2009 (Source: U.S. Department of Agriculture Forest service 2009a).

Outdoor Recreation in the Northern United States

For the North, both the total number of outdoor recreation participants and total annual activity days grew slower than the national rate. Participants increased by 3.5 percent, from about 90.4 million to 93.5 million, but their total number of annual activity days increased 23.7 percent, from 27.2 billion to 33.6 billion. Average annual activity participation days per person across the full list of activities rose from about 300 per year to 359, a 20 percent increase. Although the number of participants increased only modestly, they engaged in an average of nearly one activity per day over the course of a year (U.S. Department of Agriculture, Forest Service. 2009a). This represents a fairly significant increase in activity level compared to approximately a decade before.

The percentage of the national total number of participants and total population by region is shown for seven activity groups in Table 4. Also listed is the participation rate (percent of the region's population age 16 and older participating) for the four regions.

Table 4. Participation in seven activity groups by individuals aged 16 years and older in four U.S. regions

Activity Group (activities that comprise the group)	Region	Region's percent of U.S. participants[a]	Region's percent of U.S. population[a]	Percent of region's population participating
Visiting recreation and historic sites				
(Attending family gatherings, picnicking, visiting the beach, visiting historic or prehistoric sites, and camping)	North	42.0	40.7	82.7
	South	29.7	31.4	78.9
	Rocky Mountains	10.1	10.1	81.9
	Pacific Coast	18.2	17.8	81.4
Viewing/photographing nature				
(View/photograph birds, natural scenery, other wildlife besides birds, and wildflowers, trees, and other plants)	North	40.8	40.7	75.6
	South	30.7	31.4	73.2
	Rocky Mountains	10.5	10.1	78.1
	Pacific Coast	17.9	17.8	75.8
Backcountry activities				
(Backpacking, day hiking, horseback riding on trails, mountain climbing, and visiting a wilderness or primitive area)	North	40.1	40.7	43.1
	South	26.0	31.4	37.4
	Rocky Mountains	13.0	10.1	57.4
	Pacific Coast	20.9	17.8	51.4

Table 4. (Continued)

Activity Group (activities that comprise the group)	Region	Region's percent of U.S. participants[a]	Region's percent of U.S. population[a]	Percent of region's population participating
Motorized activities				
(Motorboating, off-highwayvehicle driving, snowmobiling, using personal watercraft, and waterskiing)	North	40.8	40.7	36.4
	South	31.1	31.4	37.1
	Rocky Mountains	10.7	10.1	39.1
	Pacific Coast	17.4	17.8	35.6
Hunting and fishing				
(Anadromous fishing, coldwater fishing, warmwater fishing, saltwater fishing, big game hunting, small game hunting, and migratory bird hunting)	North	38.6	40.7	32.4
	South	35.5	31.4	38.8
	Rocky Mountains	10.9	10.1	37.1
	Pacific Coast	15.0	17.8	28.8
Non-motorized boating				
(Canoeing, kayaking, rafting, rowing, and sailing)	North	45.6	40.7	23.0
	South	27.5	31.4	18.0
	Rocky Mountains	9.2	10.1	18.7
	Pacific Coast	17.7	17.8	20.4
Snow skiing and boarding				
(Cross country skiing, downhill skiing, and snowboarding)	North	49.6	40.7	14.0
	South	14.5	31.4	5.5
	Rocky Mountains	12.6	10.1	14.7
	Pacific Coast	23.3	17.8	15.1

[a] Percentages sum down to 100 across the four regions of each activity group. May not equal 100.0 exactly due to rounding.

Source: U.S. Department of Agriculture Forest Service 2009a.

Visiting Recreation and Historic Sites

In general, regional differences are slight with participation in activities at recreation and historic sites slightly greater in the North and slightly lower in the South. The South is the only region whose participation is less than the 81.0 percent national rate, although only about 2 percent less.

Outdoor Recreation in the Northern United States

Viewing and Photographing Nature
Participation rates are a few percentage points higher in the Rocky Mountains and Pacific Coast, and a few points lower in the South. The North participation rate of 75.6 percent is identical to the national rate.

Backcountry Activities
The percentage of people who participated in backcountry activities is substantially higher in the Rocky Mountains and Pacific Coast than in the Nation as a whole, and is especially higher than in the South. Northern participation in backcountry activities (43.1 percent) is only slightly lower than the national rate of 44.3 percent. Northerners are more likely than people in the South to be backcountry activity participants, but less likely than residents of the West.

Motorized Activities
Northern participation is just slightly lower than the national rate (36.9 percent), and lower than all other regions except the Pacific Coast. Participation in motorized activities is slightly higher in the Rocky Mountains than in the other three regions, and the Rocky Mountains is the only region that is more than a few percentage points higher than the national participation rate.

Hunting and Fishing
The North lags behind the national rate in hunting and fishing participation, by nearly 2 percent. The South leads all regions in hunting and fishing participation, followed by the Rocky Mountains. Both are higher than the national rate of 34.3 percent. Hunting and fishing participation is somewhat more likely in the North than in the Pacific Coast.

Non-Motorized Boating Activities
Participation in non-motorized boating is highest in the North and Pacific Coast, and lowest in the South. At 23.0 percent participation, the North leads the national rate of 20.8 percent and also has significantly more participation than the South and Rocky Mountains.

Snow Skiing and Boarding
Snow skiing and snowboarding participation is highest in the Pacific Coast and Rocky Mountains, followed by the North, although the rates for those three regions are separated by just 1.1 percent. Not unexpectedly,

22   H. Ken Cordell, Carter J. Betz, Shela H. Mou et al.

participation is by far lowest in the South. Every region but the South exceeds the national participation rate of 11.6 percent. Participation in the South is less than half the national rate. With about 41 percent of the national population, the North has almost half the Nation's skiers and snow boarders.

The North's Participation in Nature-Based Activities

Tables 5 through 8 summarize the trends in activity participation (number of people and percent of population age 16 and older in the North) in nature-based activities, such as birdwatching or camping, from the mid-1990s to 2009.

Activities that had more than or equal to 30 million people participating are shown in Table 5. Walking for pleasure, attending family gatherings outdoors, gardening or landscaping, viewing/photographing natural scenery, visiting outdoor nature centers or zoos, and picnicking occupied the top six slots, each with over 50 million participants in the North. In the 40-to-50-million participant category were viewing/photographing wildlife (besides birds and fish), viewing/photographing wildflowers/ trees/other plants, sightseeing, driving for pleasure, visiting a beach, visiting historic sites, swimming in lakes/ponds/streams, and swimming in an outdoor pool. With a few exceptions—visiting historic sites, picnicking, gardening or landscaping, driving for pleasure, and bicycling, which increased less than 5 percent—all of the most popular activities have shown considerable growth. Activities oriented toward viewing and photographing nature (scenery, flowers/trees/other plants, and wildlife) were among the fastest growing.

> The popularity of birding and other viewing and learning activities continues throughout the North and the United States, as illustrated by this couple and their grandsons at the Tamarac National Wildlife Refuge in Minnesota. (Photograph courtesy of Lake Country Scenic Byway Association)

Twelve activities had 10 to 30 million participants (Table 6). Viewing or photographing fish, warm water fishing, motor boating, visiting a waterside (besides a beach), sledding, and developed camping all had more than 20 million participants. Four activities—developed camping, mountain biking, primitive camping, and participating in boat tours or excursions—showed a decrease in numbers of participants during the decade. Fastest growing for this period were offhighway-vehicle driving, warm water fishing, and viewing or photographing fish.

Table 5. For activities with greater than 30 million participants annually (2005 to 2009), trends in the number and percentage of people in the North age 16 years and older participating in nature-based activities from 1994 to 2009

Activity	1994 to 1995[a]	1999 to 2001[b]	2005 to 2009[c]		1999 to 2009
	number of participants(millions)		number of participants (millions)	Portion of Population (percent)	Change (percent)
Walking for pleasure	62.9	78.1	82.5	84.6	5.6
Attending family gatherings	58.5	68.2	73.0	74.9	7.1
Gardening or landscaping	—	63.3	64.9	66.6	2.5
Viewing/photographing natural scenery	—	55.4	61.9	63.6	11.8
Visiting an outdoor nature center/zoo	50.2	53.1	55.9	57.3	5.3
Picnicking	52.1	52.6	53.2	54.6	1.0
Viewing/photographing wildlife (other than birds and fish)	29.1	41.2	49.7	51.0	20.5
Viewing/photographing flowers/trees/other plants		40.5	49.6	50.9	22.6
Sightseeing	52.6	47.2	49.5	50.8	5.0
Driving for pleasure	—	47.3	49.0	50.3	3.6
Visiting beaches	58.0	38.2	44.1	45.2	15.3
Visiting historic sites	41.1	43.4	43.3	44.4	-0.1
Swimming in lakes/ponds/streams	41.7	39.4	42.7	43.8	8.4
Swimming in outdoor pools	44.9	37.6	41.6	42.7	10.5
Bicycling	36.7	37.1	38.6	39.6	4.1
Viewing or photographing birds	25.5	31.6	37.3	38.2	17.8
Gathering mushrooms/berries	—	27.9	35.0	36.0	25.7
Visiting farm or agricultural settings	—	27.3	34.8	35.7	27.5
Day hiking	22.5	27.7	31.9	32.7	15.1
Visiting wilderness areas	—	27.5	30.5	31.3	10.7

— = Participation in this activity was not asked during this survey period.

[a] Based on regional population of 89.64 million people age 16 years and older (Woods and Poole Economics, Inc. 2009).

[b] Based on regional population of 92.43 million people age 16 years and older (U.S. Department of Commerce, Bureau of the Census 2009a).

[c] Based on regional population of 97.44 million people age 16 years and older (U.S. Department of Commerce, Bureau of the Census 2009a).

Source: U.S. Department of Agriculture Forest Service 2009a.

Table 6. For activities with 10 to 30 million participants (2005 to 2009), trends in the number and percentage of people in the North age 16 years and older participating in nature-based activities from 1994 to 2009

Activity	1994 to 1995[a]	1999 to 2001[b]	2005 to 2009[c]		1999 to 2009
	number of participants (millions)		number of participants (millions)	Portion of Population (percent)	Change (percent)
Viewing/photographing fish	11.6	21.7	24.6	25.2	13.1
Warmwater fishing	22.9	20.4	23.9	24.5	17.3
Motorboating	27.1	22.5	23.5	24.1	4.7
Visiting watersides (besides beaches)	—	22.6	23.0	23.6	1.9
Sledding	17.7	19.8	20.7	21.3	4.5
Developed camping	19.6	22.4	20.1	20.6	-10.4
Mountain biking		21.2	19.8	20.3	-6.4
Participating in boat tours or excursions	—	19.1	18.7	19.2	-1.9
Visiting prehistoric sites	14.9	17.6	18.1	18.6	3.0
Driving off-road	14.0	13.8	17.2	17.6	24.8
Canoeing	9.8	11.1	12.0	12.3	8.2
Primitive camping	11.8	11.9	11.6	11.9	-2.5

— = Participation in this activity was not asked during this survey period.

[a] Based on regional population of 89.64 million people age 16 years and older (Woods and Poole Economics, Inc. 2009).

[b] Based on regional population of 92.43 million people age 16 years and older (U.S. Department of Commerce, Bureau of the Census 2009a). [c] Based on regional population of 97.44 million people age 16 years and older (U.S. Department of Commerce, Bureau of the Census 2009a).

Source: U.S. Department of Agriculture Forest Service 2009a.

Among the 23 activities with 3 to 10 million participants (Table 7), big game hunting, backpacking, ice skating outdoors, saltwater fishing, and waterskiing were at the top of the list. Seven of the activities posted double-digit percentage growth since 1999; eight activities grew from 0 to 10 percent; and eight activities decreased in number of participants. The fastest growing activities were kayaking (which grew at more than twice the rate of any other activity in this category), snowboarding, and waterskiing. Cross-country skiing, downhill skiing, and rafting posted the largest percent decreases.

Six activities had fewer than 3 million participants (Table 8). At the top of the list, with 2 or more million participants, were orienteering (which grew

Outdoor Recreation in the Northern United States 25

nearly 91 percent since 1999) and snowshoeing (which decreased more than 17 percent). Given their low participation rates, these activities primarily represent niche markets that appeal to small population segments. Many require substantial investments in time, equipment, and skill.

The participation data shown in Tables 5 through 8 in part may be reflecting the rapid rise in gasoline prices from 2007 to 2008 and the recession that began in 2007. Viewed overall, however, it is clear that what people in the North choose as activities is changing over time. Some of the activities that dominated a generation or two ago no longer dominate with the emergence of underlying changes in society, generations, lifestyles, information, and technology (Cordell 2008).

A timely topic that has captured the attention of many conservation leaders and other interested supporters throughout the Nation is the relationship that America's youth has with natural resources. In particular, many observers have expressed concern over what they see as a growing "disconnect" between children and the outdoors, asserting that children in America are spending increasingly less time outdoors. The following section presents results of a study of their time outdoors that was conducted from 2007 to 2009.

Children and the Outdoors

Better understanding of outdoor time and activities among children provides some very important insights into the future. To address this need for better understanding, data from the National Kids Survey were analyzed to estimate the portion of each day that 6-to19-year-olds spent outdoors during the week preceding their interviews (Cordell 2012). This survey is the only national data source of time and activities by young people outdoors.

Estimates from the kids survey included outdoor time during a typical weekday and typical weekend day (Table 9). As well, current estimates of time outdoors per day are compared to the previous year (Table 10). Nationally, about 62 percent reported spending two or more hours outdoors per day on a typical weekday and 77 percent on typical weekend days, compared to 58 percent on weekdays and 74 percent on weekends in the North (Table 9). Just under half of youths nationwide spent 4 or more hours outdoors on a typical weekend day, compared to 43 percent of North region youths. Less than 5 percent spent no time outdoors on either weekdays or weekend days regardless of where the youth lived. As one might expect, school and other activities that

26 H. Ken Cordell, Carter J. Betz, Shela H. Mou et al.

are not necessarily recreation likely compose a significant amount of youth time outdoors during weekdays.

Table 7. For activities with 3 to 10 million participants (2005 to 2009), trends in the number and percentage of people in the North age 16 years and older participating in nature-based activities from 1994 to 2009

Activity	1994 to 1995[a]	1999 to 2001[b]	2005 to 2009[c]		1999 to 2009
	number of participants- (millions)		number of participants (millions)	Portion of Population (percent)	Change (percent)
Big game hunting	8.6	7.3	8.5	8.7	17.1
Backpacking	6.5	8.5	8.3	8.5	-1.6
Ice skating	10.5	9.1	8.2	8.4	-9.7
Saltwater fishing	8.1	7.9	8.0	8.2	2.0
Waterskiing	9.3	5.9	7.6	7.8	29.2
Horseback riding	7.9	7.1	7.6	7.8	6.1
Use personal watercraft	4.5	7.1	7.4	7.6	3.8
Downhill skiing	11.0	8.7	7.3	7.5	-15.2
Rafting	8.4	8.1	7.0	7.2	-13.2
Snowmobiling	6.5	7.1	6.9	7.1	-2.7
Kayaking	1.4	3.6	6.8	7.0	89.3
Small game hunting	7.5	5.7	6.6	6.8	15.4
Snorkeling[d]	6.4	5.6	5.8	5.9	3.9
Horseback riding on trails	6.1	5.8	5.8	5.9	0.1
Snowboarding	3.3	4.1	5.6	5.8	35.9
Rowing	6.4	4.8	4.9	5.0	1.4
Sailing	5.8	5.5	4.9	5.0	-11.6
Mountain climbing	3.0	4.6	4.1	4.2	-11.6
Caving	3.9	3.1	4.0	4.1	30.0
Rock climbing	2.9	3.5	3.9	4.0	10.0
Cross country skiing	5.8	4.8	3.9	4.0	-20.6
Ice fishing	3.7	3.5	3.8	3.9	9.2
Anadromous fishing	5.0	3.4	3.6	3.7	6.2

[a]Based on regional population of 89.6 million people age 16 years and older (Woods and Poole Economics, Inc. 2009).

[b]Based on regional population of 92.4 million people age 16 years and older (U.S. Department of Commerce, Bureau of the Census 2009a). [c]Based on regional population of 97.4 million people age 16 years and older (U.S. Department of Commerce, Bureau of the Census 2009a). [d]Scuba diving was included in the snorkeling activity in this survey period.

Source: U.S. Department of Agriculture Forest Service 2009a.

Outdoor Recreation in the Northern United States — 27

Table 8. For activities with less than 3 million participants (2005 to 2009), trends in the number and percentage of people in the North age 16 years and older participating in nature-based activities from 1994 to 2009

Activity	1994 to 1995[a]	1999 to 2001[b]	2005 to 2009[c]		1999 to 2009
	number of participants (millions)		number of participants (millions)	Portion of Population (percent)	Change (percent)
Orienteering	2.1	1.3	2.4	2.5	90.6
Snowshoeing	—[d]	2.7	2.3	2.3	-17.1
Migratory bird hunting	2.0	1.6	1.8	1.8	12.9
Scuba diving	—[e]	1.4	1.5	1.5	6.7
Surfing	0.8	0.8	1.1	1.1	43.4
Windsurfing	1.2	0.7	0.8	0.8	3.7

— = Missing data.

[a]Based on regional population of 89.64 million people age 16 years and older (Woods and Poole Economics, Inc. 2009).

[b]Based on regional population of 92.43 million people age 16 years and older (U.S. Department of Commerce, Bureau of the Census 2009a). [c]Based on regional population of 97.44 million people age 16 years and older (U.S. Department of Commerce, Bureau of the Census 2009a). [d]Participation in this activity was not asked during this survey period.

[e]Scuba diving was included as part of snorkeling in 1994 to 1995.

Source: U.S. Department of Agriculture Forest Service 2009a.

Table 9. For 6- to 19-year-olds in the United States and Northern States, time spent outdoors on typical weekdays and weekend days

	Respondents (percent)	
Amount of time	On weekdays	On weekend days
United States		
None	2.3	3.9
<1/2 hour a day	4.3	2.2
About 1/2 hour a day	8.1	3.5
About 1 hour	23.1	13.3
2-3 hours	33.8	27.3
≥4 hours	28.5	49.8
Northern States		
None	3.3	4.5
<1/2 hour a day	4.0	1.8
About 1/2 hour a day	9.2	3.8

Table 9. (Continued)

	Respondents (percent)	
Amount of time	On weekdays	On weekend days
About 1 hour	25.5	15.7
2-3 hours	32.2	31.1
≥4 hours	25.9	43.1

Note: Percent may not sum down within the United States and North to 100.0 exactly due to rounding.
Source: Larson et al. (2011).

Table 10. For 6- to 19-year-olds in the United States and Northern States, amount of time spent outdoors compared to the same time last year

Sample	Less time	About the same	More time
United States	15.5	44.9	39.6
Northern States	17.0	49.0	34.0

Note: Percent may not sum across within the United States and North to 100.0 exactly due to rounding.
Source: Larson et al. 2011.

Next examined were percentages who indicated spending less, the same or more time outdoors at the time of the interview relative to a year ago. Across the entire sample, both boys and girls, only about 16 percent reported spending less time, 45 percent reported spending the same, and 40 percent estimated spending more time outdoors this year than last (Table 10). In the North, respondents were slightly more likely to report spending less time outdoors than they did a year ago (17 percent), more likely to say they spent about the same amount of time outdoors (49 percent), and less likely to say they spent more time outdoors (34 percent).

Table 11 compares outdoor activity participation rates (percentages) between male and female respondents in the North and the Nation. Playing outdoors or "hanging out" during the previous week was the most common activity, with about 84 percent participating.

Male participation in this unstructured free play was higher both nationwide and in the North. The grouping of biking/jogging/walking/ skateboarding or similar activities was the next most popular with nearly 80 percent participating both nationally and in the North. For both samples, female participation was slightly higher than male participation. Listening to music or using a screen or other electronic device outdoors was the third most

Outdoor Recreation in the Northern United States 29

cited outdoor activity with about 51 percent participating. This activity was slightly more popular with girls than with boys. These data indicate that the use of electronic media is not limited to indoor settings. Nationally, playing or practicing team sports was considerably more popular with boys, but less so in the North. By contrast, reading or studying while sitting outdoors was more popular with girls, both nationally and in the North.

Many children and teenagers enjoy spending time outdoors in nature-based recreation as well as in other activities closer to home. Here, children learn how to kayak in the Hiawatha National Forest in Michigan. (Photograph by Anne Okonek)

Table 11. For 6- to 19-year-olds in the United States and Northern States, participation (percent) in outdoor activities during the past week, by gender

Outdoor activities	United States			Northern States		
	Male	Female	Total	Male	Female	Total
Just playing outdoors or hanging out	87.1	81.8	84.5	87.9	79.4	83.9
Biking/jogging/walking/skateboarding	78.3	81.4	79.7	75.8	82.4	78.8
Listening to music or using other electronic devices	48.2	55.2	51.6	48.5	53.8	51.0
Playing or practicing team sports	59.5	38.8	49.3	54.7	43.8	49.5
Reading or studying while sitting outdoors	38.5	53.5	45.8	36.8	52.8	44.2
Participating in individual sports, (such as tennis, golf)	38.0	35.2	36.6	34.7	38.4	36.5
Attending camps, field trips, outdoor classes	34.1	39.1	36.5	36.1	37.1	36.3
Swimming, diving, snorkeling	30.8	32.1	31.5	32.5	24.4	28.7
Birdwatching and wildlife viewing	28.5	32.3	30.4	30.2	27.4	28.9
Hiking, camping, fishing	30.8	28.2	29.5	36.1	24.9	30.9
Riding motorcycles, all-terrain vehicles, other off-road vehicles	24.2	14.4	19.4	23.7	13.7	19.0
Boating, jet skiing, water skiing	9.1	7.2	8.2	8.4	4.5	6.5
Snow skiing, snowboarding, cross-country skiing	8.6	6.6	7.6	12.2	8.7	10.6
Rowing, kayaking, canoeing, surfing	8.1	6.8	7.5	7.3	7.4	7.4
Participating in other activities	10.3	11.6	10.9	13.8	8.6	11.3

Note: Each activity asked separately; sample sizes vary.
Source: Larson et al. 2011

More than a third of respondents participated in other sports such as golf and tennis and in organized nature-based activities such as attending outdoor camps, classes, and field trips. These two activity groups had nearly identical participation rates of 36 percent, both nationally and in the North. Gender differences were slight in both samples. In the North, a number of other nature-based activities (swimming, diving, snorkeling, birdwatching, wildlife viewing, and hiking, camping, fishing) each experienced about 30 percent participation. Boys outpaced girls in the motorized activity of riding off-road vehicles by about 10 percent both nationally and in the North. Hiking, camping, and fishing participation also was higher for males than females in the North.

Table 12 shows participation by age group for the same set of activities in the North and for the United States In both samples, unstructured free play decreased with age; biking/jogging/ walking/skateboarding decreased up to age 15 and then rebounded slightly with the oldest age group of 16- to-19-year-olds. Electronic media use outdoors is much less frequent with the youngest age group, but then rises with age until peaking with the 13- to-15-year-age group.

In both the North and the Nation, team sports also peaked with the early teenage group, but participation in reading or studying outdoors was considerably higher for the oldest teens than for the three younger groups. Other, mostly individual sports, ranked higher in the 10- to12-year-old group and also in the youngest group, relative to the two older ones. This may be indicative of more children beginning these activities at an early age then dropping out to pursue other interests as they get older. Interestingly, activities such as attending outdoor classes and camps was most popular nationwide with the youngest age group, but most popular in the North with the oldest age group. This may be because the northeastern area has a stronger tradition of organized camps and more organized camping facilities than other parts of the country. Participation rates for these activities were fairly consistently in the 33-to-40-percent range regardless of age or location. Birdwatching and wildlife viewing decreased significantly with age in both samples, which may be an indicator of family outings in which teenagers do not participate. Off-road vehicle driving and boating activities increased with age in both samples, although boating participation peaked with early teens nationally and with 10-to-12-year-olds in the North.

Outdoor Recreation in the Northern United States 31

Table 12. For 6- to 19-year-olds in the United States and Northern States, participation (percent) in outdoor activities during the past week, by age group

Outdoor activities	United States				Northern States			
	Age 6 to 9	Age 10 to 12	Age 13 to 15	Age 16 to 19	Age 6 to 9	Age 10 to 12	Age 13 to 15	Age 16 to 19
Just playing outdoors or hanging out	91.5	95.4	82.8	69.8	90.2	93.2	82.9	71.6
Biking/jogging/walking/ skateboarding	85.1	82.8	70.9	78.7	82.1	79.4	74.4	78.5
Listening to music, watching movies, using electronics	33.4	50.0	63.8	63.8	34.7	51.4	59.6	59.1
Playing or practicing team sports	45.1	49.4	56.5	48.6	43.3	51.3	56.2	50.0
Reading or studying while sitting outdoors	42.1	47.0	36.2	56.3	41.2	39.1	37.0	56.9
Participating in individual sports (such as tennis, golf)	41.7	45.0	28.1	31.3	38.2	44.5	31.3	33.1
Attending camps, field trips, outdoor classes	40.6	37.3	33.2	34.0	36.4	35.5	34.5	39.1
Swimming, diving, snorkeling	31.5	32.2	31.5	30.6	27.4	29.3	29.0	29.1
Birdwatching and wildlife viewing	40.9	36.8	23.8	19.3	35.9	38.9	29.8	14.0
Hiking, camping, fishing	34.3	26.2	27.5	28.6	32.2	34.3	29.0	28.6
Riding motorcycles, all-terrain vehicles, other off-road driving	17.8	13.2	22.1	24.1	16.7	18.0	20.4	21.2
Boating, jet skiing, water skiing	6.7	7.3	13.0	9.9	3.4	8.0	6.7	8.7
Snow skiing, snowboarding, cross-country skiing	7.0	8.2	8.9	8.7	10.1	16.7	6.8	10.0
Rowing, kayaking, canoeing, surfing	5.7	8.4	6.2	10.2	4.1	9.9	5.8	10.4
Participating in other activities	9.7	11.3	5.4	10.4	7.8	10.9	16.4	10.8

Note: Each activity asked separately; sample sizes vary. Source: Larson et al. 2011

RECREATION RESOURCE TRENDS AND FUTURES

Federal Recreation Resources

Federal Land

Nationally, Federal agencies manage nearly 640 million acres, much of which includes vast areas suitable for a variety of outdoor recreation activities. Such areas are as important in the North as they are throughout the country. Other than some national wildlife refuges areas reserved for science and research, dams, and other administrative and operational sites, very little Federal land is closed or has restrictions on public access, but access is sometimes blocked by in-holdings and ownership fragmentation.

Less than 3 percent of Federal land (about 17.9 million acres) is in the North, about 69 percent of which is managed by the Forest Service. More than 92 percent of Federal land is located in the western United States. Even not counting Alaska, which has 36 percent of the national total, Federal land is predominantly western, making up 88 percent of the 49-State total area. The regional distribution of acreage in the three water resources agencies (U.S. Army Corps of Engineers, U.S. Department of the Interior Bureau of Reclamation, and Tennessee Valley Authority), however, is much more evenly distributed between the West and East. Of the total land and water area in these three agencies, nearly half is located in the East, about 12 percent in the North and 37 percent in the South.

Federal acreage changes very little over time. What does change, however, particularly by region, is the amount of Federal land per capita as population grows. In 2008, the 2,105 acres per 1,000 U.S. residents (or about 2.1 acres per person) represented a 5.6 percent decrease from the 2002 level. Decreases were largest in the Rocky Mountains (8.8 percent) and Pacific Coast (7.7 percent), reflecting greater population growth in those regions. The North, with 143.6 Federal acres per 1,000 persons, had the smallest regional decrease (-2.4 percent).

The decrease in per capita Federal acres nationally was even more pronounced when compared to 1995 levels, mirroring the 14 percent population increase (Table 13). The 8.2 percent decrease in Federal acres per capita in the North was the slowest of any region and considerably less than the national rate of change. Nonetheless, pressures for recreation space in the North are likely to increase as population grows, albeit more slowly than in the past.

Outdoor Recreation in the Northern United States

Wilderness

The North accounts for just 1.5 percent, or about 1.7 million of the 109.5 million acres in the National Wilderness Preservation System. Similar to Federal land in general, the Wilderness land managed by four Federal agencies lies mostly in Western States (96 percent). Alaska, in particular, has more than 52 percent of the total system—largely managed by the U.S. Department of the Interior National Park Service and U.S. Department of the Interior Fish and Wildlife Service.

Table 13. Federal acres per 1,000 people (including Alaska) in 1995 and percentage change from 1995 to 2008, by region; estimated U.S. population was 266.28 million (Woods and Poole Economics, Inc. 2009) for 1995 and 304.06 million (U.S. Department of Commerce, Bureau of the Census 2009a) for 2008

Agency	North		South		Rocky Mountains		Pacific Coast		United States	
	Acres 1995[a]	Percent Change 2008	Acres 1995[a]	Percent Change 2008	Acres 1995[a]	Percent Change 2008	Acres 1995[a]	Percent Change 2008	Acres 1995[a]	Percent Change 2008
Forest Service	101.9	-3.4	151.7	-14.6	4,600.3	-22.3	1,581.1	-12.7	719.6	-11.9
National Park Service	11.0	-1.8	58.3	-13.4	482.5	-17.5	1,447.3	-13.8	292.0	-11.2
Fish and Wildlife Service	10.3	34.0	44.8	-5.4	330.6	7.6	1,855.6	-13.7	339.7	-8.5
Bureau of Reclamation	0.0	0.0	2.3	-17.4	251.4	-21.8	20.3	-14.3	24.5	-12.7
Bureau of Land Management	3.3	-100.0	9.4	-95.7	6,629.1	-22.5	2,898.7	-22.4	1,005.1	-17.1
Tennessee Valley Authority	0.0	0.0	2.9	-17.2	0.0	0.0	0.0	0.0	0.9	-11.1
Army Corps of Engineers	24.8	-16.9	66.2	4.4	113.8	11.9	12.8	-13.3	43.4	4.1
All agencies	156.4	-8.2	350.2	-15.4	12,422.9	-21.2	7,911.8	-17.8	2,448.6	-14.0

[a]Resource data years for earlier period vary by agency; expressed as 1995 because 1995 population estimates were used in per capita measures.

Sources: U.S. Department of Agriculture Forest Service 1995, 2008; U.S. Department of the Interior National Park Service 1995, 2008; U.S. Department of the Interior Fish and Wildlife Service 1995, 2008; U.S. Department of the Interior Bureau of Reclamation 1993, 2008; U.S. Department of the Interior Bureau of Land Management 1994, 2008; Tennessee Valley Authority 2008; U.S. Army Corps of Engineers 2006.

34 H. Ken Cordell, Carter J. Betz, Shela H. Mou et al.

Without the Alaska acres, the North's share of wilderness rises only to 3.2 percent of the Nation's total.

Since 1995, wilderness system area has grown about 6 percent, but population increases have reduced the national per capita acres by 3 percent (Table 14). In the South, the decrease was nearly 16 percent, followed by 10 percent for Oregon, California, and Washington on the Pacific Coast, and 8 percent for the Rocky Mountains. Only the North experienced an increase since 1995, although at 1.5 percent, just slightly. In the 49 States, per capita wilderness acres decreased across all agencies, except for the U.S. Department of the Interior Bureau of Land Management. All of the wilderness system acreage added in the North region is managed by the National Park Service.

Table 14. Federal acres in the National Wilderness Preservation System (excluding Alaska) per 1,000 people in 1995[a] and percentage change from 1995 to 2009[b], by region; estimated U.S. population was 265.67 million, excluding Alaska (Woods and Poole Economics, Inc. 2009) for 1995 and 303.37 million, excluding Alaska (U.S. Department of Commerce, Bureau of the Census 2009a) for 2008

Agency	North		South		Rocky Mountains		Pacific Coast		United States	
	Acres 1995	Percent Change 2009	Acres 1995	Percent Change 2009	Acres 1995	Percent Change 2009	Acres 1995	Percent Change 2009	Acres 1995	Percent Change 2009
Bureau of Land Management[a]	0.0	0.0	0.0	0.0	74.8	121.4	89.5	-3.5	20.0	44.0
Fish and Wildlife Service[a]	0.5	0.0	5.5	-16.4	67.3	-21.7	0.3	-33.3	7.6	-11.8
Forest Service[b]	11.5	0.0	8.3	-12.0	823.1	-20.5	237.8	-9.1	111.5	-9.5
National Park Service[a]	1.1	27. 3	17.5	-17.1	36.0	34.2	200.9	-13.8	39.0	-6.9
U.S. Total	13.2	1.5	31.3	-15.7	1,001.2	-8.0	505.2	-9.8	176.9	-3.0

[a]U.S. Department of the Interior
[b]U.S. Department of Agriculture
Source: Wilderness.net 2009.

Protected Rivers and Trails

Two Federal systems play a key role in resource protection and outdoor recreation. They are the National Wild and Scenic Rivers System and the National Recreation Trails System, both established by Congress in 1968. The currently more than 12,500 miles of wild and scenic rivers in the United States represent an 11 percent increase since 2000 (Table 15); 3,000 miles are in the East and the remaining 76 percent are in the West. These rivers—which are classified as wild, scenic, or recreational—range from the most primitive and undeveloped (wild) to the most accessible and (perhaps) impounded in the past (recreational). The North has nearly 2,200 miles (about 17 percent of the national total), an increase of 6 percent since 2000. Most of the 125 miles added in the North are in the scenic and recreational classifications, with only 2 miles in the wild classification.

The National Trails System consists of three categories of nationally significant trails: National Scenic Trails, National Historic Trails, and National Recreation Trails. Similar to the federally designated rivers, national trails protect linear land resources that are judged to have significant value for the entire country.

The scenic and historic category typically consists of long overland trails that are remote from population centers, compared to recreation trails, which tend to be located near or within urban areas. As of 2009 the United States had more than 1,000 national recreation trails totaling more than 20,000 miles (Table 16); of these, 53 percent of the trails and nearly 69 percent of the mileage are located in the populous East. The North, which had more than 7,300 miles and 36 percent of the system, has added more trails than any other region and more miles of trails (3,200) than any other region except the South, which saw 84 percent growth in trail miles since 2004.

Recreation Facilities

The Recreation Information Database, an interagency effort coordinated by the U.S. Department of the Interior, is a public data portal on Federal recreation sites and facilities throughout the country. Table 17 shows that the Nation's estimated 9,075 Federal facilities translate into just under 30 facilities per million people (or about 1 per 33,500). With just 9.5 facilities per million people overall (or about 1 per 105,000), the North lags behind the western regions by a wide margin. The combination of much more Federal property and much less population in the West is a primary reason for this disparity. The North also lags behind the South, which has more camping and boating facilities but is about even in other facilities. The Rocky Mountains has more than 10 times the number of available Federal facilities per capita than both the South and the North and has nearly twice as many as the Pacific Coast. Camping facilities dominate in the listing of facilities being offered at nearly 96 percent of areas nationwide.

Table 15. Miles of river in the National Wild and Scenic River System by classification and region, 2000 and 2009, with percentage change (includes AK and HI)

	Wild rivers			Scenic rivers			Recreational rivers			Total		
	2000 (miles)	2009 (miles)	Percent Change	2000 (miles)	2009 (miles)	Percent Change	2000 (miles)	2009 (miles)	Percent Change	2000 (miles)	2009 (miles)	Percent Change
North	172	174	1.5	935	1,014	8.5	964	1,007	4.4	2,070	2,195	6.0
South	187	284	51.8	318	414	30.2	112	112	0.0	617	810	31.3
Rocky Mountains	710	1,328	87.1	288	380	31.9	532	587	10.5	1,530	2,295	50.0
Pacific Coast	4,280	4,370	2.1	911	936	2.7	1,886	1,946	3.2	7,077	7,252	2.5
U.S. total	5,349	6,156	15.1	2,452	2,743	11.9	3,493	3,652	4.6	11,294	12,552	11.1

Source: Interagency Wild and Scenic Rivers Council 2009.

Table 16. Number and miles of National Recreation Trails by region, 2004 and 2009, with percentage change (includes AK and HI)

	National Recreation Trails					
Region	Number		Percent	Miles		Percent
	2004	2009	Change	2004	2009	Change
North	226	312	38.1	4,119	7,319	77.7
South	220	264	20.0	3,578	6,577	83.8
Rocky Mountains	254	292	15.0	2,969	3,380	13.8
Pacific Coast	198	209	5.6	2,622	2,944	12.3
U.S. total	898	1,077	19.9	13,288	20,220	52.2

Source: American Trails 2010.

Non-Federal Recreational Resources

State Parks

Each of the 50 States has a State park system, which is usually a division or agency within the department of natural resources or conservation. These resources are usually closer to population centers and more developed than their Federal counterparts. Although most State park systems manage a significant number of backcountry acres, remote holdings are not nearly as common as they are in Federal systems. State parks have been called "intermediate" resources because they represent a middle ground between the sometimes vast and distant Federal lands and the usually much smaller and more highly developed parks managed by local governments (Clawson and Knetsch 1966).

State systems predominantly feature State parks, but they also include facilities classified as recreation areas, natural areas, historic sites, environmental education and science areas, State forests, and wildlife and fish management areas. Although these other types of protected areas may exist separately from State park systems, they are usually still housed within natural resource departments, as are most—but not all— wildlife and fish areas and State forests.

Table 17. Federal recreation facilities provided or activities permitted per million people by region, 2009; based U.S. population estimate of 304.06 million (U.S. Department of Commerce, Bureau of the Census 2009a) in 2008 (includes AK and HI)

Activity or facility	North	South	Rocky Mountains	Pacific Coast	United States
	Number available (per million people)[a]				
Camping	8.3	11.2	121.3	63.8	28.6
Hiking	1.5	1.8	65.6	19.9	10.4
Fishing	1.3	2.4	64.0	18.3	10.2
Boating	1.9	4.3	22.7	10.2	5.9
Picnicking	0.1	0.1	43.6	9.3	5.6
Recreational vehicle camping	0.0	0.0	38.0	11.7	5.4
Biking	0.4	0.4	32.4	5.7	4.2
Horseback riding	0.1	0.4	27.5	4.7	3.5
Hunting	0.4	0.8	24.8	4.2	3.4
Wildlife viewing	0.1	0.1	20.1	7.5	3.1
Auto touring	0.0	0.0	13.4	2.4	1.6
Water sports	0.0	0.0	6.4	3.9	1.2
Interpretive programs	0.8	0.7	4.7	1.3	1.2
Visitor centers	0.9	0.8	4.0	1.1	1.2
Riding off highway vehicles	0.0	0.0	9.3	1.2	1.0
Wildernesses	0.0	0.0	6.0	2.3	0.9
Winter sports	0.0	0.0	6.3	0.8	0.7
Swimming sites	0.0	0.0	2.2	2.8	0.6
Historic and cultural sites	0.2	0.0	4.0	0.5	0.5
Fish hatcheries	0.1	0.2	0.7	0.3	0.2
Day use areas	0.0	0.0	1.4	0.4	0.2
Climbing	0.0	0.0	1.5	0.2	0.2
All activities and facilities	9.5	12.1	124.2	65.2	29.8

Note: Activities shown as 0.0 reflect <0.05 per million people.
Source: U.S. Department of the Interior 2009.

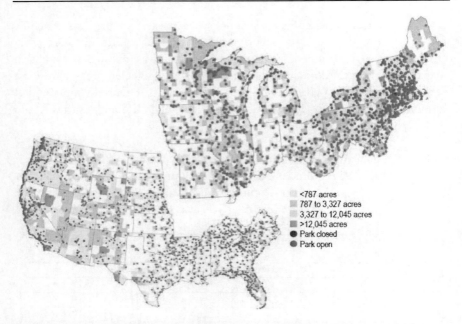

Figure 9. Location and status of State park system units and acres by county in the contiguous United States, 2009 (Source: U.S. Department of Agriculture Forest Service 2009b).

State park systems are usually accessible to the general public as is evidenced by their distribution across U.S. counties (Fig. 9), the majority of which have some acreage of State park system lands. Numerous State park system areas are evident throughout much of the northeastern and midwestern areas, Florida, and along the Pacific Coast. Although there are many fewer state parks in the West, they tend to be large. And although many of the largest properties are found in the West, State park system resources are also numerous throughout the East and particularly on the northern seaboard. With the exception of some parts of the Great Plains and a few other scattered regions across the country, it is rare to travel across more than just a few counties without encountering State park system lands.

Because State parks typically provide a diversity of recreation opportunities, many of the activities that people enjoy on Federal lands can also be enjoyed on the State park system lands. In the North, a State park is located within an hour's drive regardless of where one is located; the few exceptions to this rule are mostly in remote areas of northern Maine and upstate New York.

The Adirondack and Catskill Forest Preserves in New York, although not technically classified as State parks, encompass nearly 3 million acres of State-owned public land and provide numerous outdoor recreation opportunities. Furthermore, Maine, New Hampshire, Vermont, and New York all have a longstanding tradition of public access to private forest lands, particularly on parcels of 1,000 acres or more (Daigle et al. 2012).

The National Association of State Park Directors reported in their 2009 Annual Information Exchange that more than 13.9 million acres exist in State park systems, an increase of about 3 percent in acres per 1,000 people since 1995 (Table 18). Northern States reported about 5.2 million acres (37 percent of the national total or nearly 49 percent if Alaska's large State parks are removed from the Pacific Coast total). State park acreage per capita fell 2 percent from 1995 to 2008 in the North. However, State recreation area acreage per capita increased 50 percent in the North during this period. The North growth in per capita acres across all categories of areas under State park system management was more than 33 percent, about 10 times that of the Nation. It should be noted, however, that most of the North's increase was likely due to the reclassification of other State properties into the State park system's jurisdiction. This was particularly the case in New York State, which included data on the Forest Preserves and other properties managed by the Department of Environmental Conservation in the Annual Information Exchange report.

State park systems have faced difficult budgetary pressures since the onset of the 2007 recession. There have been occasional closures of some parks (for example, four in Arizona in 2010), transfers of some to other government and quasi-government entities, and reduced hours, services, and staffing (Table 19). The two affected States located in the North were Massachusetts and Michigan.

State Facilities

Table 20 shows the eight major types of facilities provided by State park systems and trends in these facilities since 1995. Of these, campsites are by far the most numerous. Nationally, access to improved (or developed) campsites, cabins, golf courses, and marinas held steady since 1995, but fell by about 12 percent for primitive campsites. In the North, the decrease in primitive campsites per capita was particularly sharp, falling by about 31 percent. Improved campsites, however, increased by 15 percent; when combined with losses in the South and Pacific Coast, the result was little to no net change nationwide.

40 H. Ken Cordell, Carter J. Betz, Shela H. Mou et al.

The drop in the number of swimming pools per capita at Northern State parks was in the same direction as the national trend. The region also experienced a reduction in the number of stables per capita, but the base year number in 1995 was already relatively small.

Table 18. State park system area per 1,000 people in 1995 and percentage change from 1995 to 2008, by region; estimated U.S. population was 266.28 million (Woods and Poole Economics, Inc. 2009) for 1995 and 304.06 million (U.S. Department of Commerce, Bureau of the Census 2009a) for 2008 (includes AK and HI)

Type Type	North		South		Rocky Mountains		Pacific Coast		United States	
	Acres 1995	Percent Change 2008	Acres 1995	Percent Change 2008	Acres 1995	Percent Change 2008	Acres 1995	Percent Change 2008	Acres 1995	Percent Change 2008
State parks	18.0	-2.3	10.1	40.2	37.0	-12.8	95.3	-8.1	29.4	-1.1
Recreation areas	1.5	49.7	1.3	-28.8	8.9	-22.9	18.1	-28.1	4.7	-15.0
Historic sites	0.1	123.1	0.3	-8.8	1.2	-59.8	0.4	51.2	0.3	12.1
Natural areas[a]	0.9	77.6	0.1	9066.7[b]	0.3	2907.4[b]	0.0	.	0.4	804.9
Other areas[c]	9.4	111.6	0.6	-7.9	14.4	-87.1	1.2	7.5	5.7	52.5
All areas	31.2	33.2	19.7	8.8	70.5	-29.7	117.4	-10.1	44.4	3.2

[a]Includes environmental education sites and areas classified as scientific sites. Large changes are likely to be the result of system reclassifications and not additions.
[b]Very large percentage change is primarily the result of system reclassifications and not additions. [c]Includes forests, fish and wildlife areas, and other miscellaneous State park system sites.
Source: National Association of State Park Directors 1996, 2009.

Local Governments

The 2007 Census of Governments tallied 8,852 local governments that provide recreation and park services, with more than 48 percent of these (4,273 units) in the North. On a proportional basis (per million people) the North experienced about 14 percent growth since 1997, just slightly higher than the national rate of 13 percent (Table 21).

Municipal recreation departments grew much faster (23 percent) than county departments and at about twice the rate of township agencies. Special recreation and park districts were the only type of local government jurisdiction that experienced reductions since 1997.

Outdoor Recreation in the Northern United States

Table 19. State park systems affected by closure or reduction in services as of 2009

State	Total facilities	Number of closures	Reduction in services
Alabama	23	None	One park transferred to county government.
Arizona	28	Two parks and two historic sites	Reduced hours for two State parks and five historic parks.
Georgia	63	None	One park changed to outdoor recreation area; reduced. hours for six historic parks/sites; and three historic sites now operated by the counties within which they reside.
Hawaii	50	None	One park transferred to a development corporation.
Massachusetts	136	Two State forests	Staffing eliminated for two areas.
Michigan	93	None	Reduced summer hours for one site.

Source: U.S. Department of Agriculture Forest Service 2009b.

Table 20. State park system area per 1,000 people in 1995 and percentage change from 1995 to 2008, by region; estimated U.S. population was 266.28 million (Woods and Poole Economics, Inc. 2009) for 1995 and 304.06 million (U.S. Department of Commerce, Bureau of the Census 2009a) for 2008 (includes AK and HI)

Facility	North		South		Rocky Mountains		Pacific Coast		United States	
	Acres 1995	Percent Change 2008	Acres 1995	Percent Change 2008	Acres 1995	Percent Change 2008	Acres 1995	Percent Change 2008	Acres 1995	Percent Change 2008
Improved campsites	608.1	15.0	361.1	-2.4	837.2	4.7	514.8	-40.4	533.2	0.3
Primitive campsites	144.4	-31.3	60.6	28.7	855.1	-5.4	215.0	-30.5	186.9	-11.7
Cabins	23.3	11.7	30.1	-11.2	17.3	50.4	9.7	29.0	22.8	5.5
Golf courses	0.4	19.5	0.6	-3.2	0.2	38.9	0.1	-42.9	0.4	4.8
Golf holes	6.4	28.2	9.8	3.3	2.5	108.9	1.7	-46.2	6.4	15.1
Marinas	0.8	29.8	1.0	-22.9	2.5	-10.7	0.3	0.0	0.9	3.3
Swimming pools	1.4	-8.1	1.5	-19.9	0.4	8.1	0.1	100.0	1.1	-12.6
Stables	0.3	-10.0	0.3	92.6	0.6	-40.0	0.1	-77.8	0.3	14.3

Source: National Association of State Park Directors 1996, 2009).

Table 21. Number of local government parks and recreation departments per million people in 1997 and 2007, and percentage change from 1997 to 2007; estimated U.S. population was 272.65 million (Woods and Poole Economics, Inc. 2009) for 1997 and 301.29 million (U.S. Department of Commerce, Bureau of the Census 2009a) for 2009 (includes AK and HI)

	North			South			Rocky Mountains			Pacific Coast			United States		
	1997	2007	Percent Change	1997	2007	Percent Change	1997	2007	Percent Change	1997	2007	Percent Change	1997	2007	Percent Change
County	3.5	3.5	0.3	5.4	5.7	6.2	5.8	5.1	-12.6	2.3	2.2	-5.6	4.1	4.2	1.7
Municipal	15.0	18.4	22.7	15.5	18.8	20.7	21.0	29.4	40.3	12.9	14.0	8.0	15.3	18.8	22.6
Town or Township	8.5	9.4	11.0	0.0	0.0	0.0	0.0	0.2	0.0	0.0	0.0	0.0	3.7	3.9	5.1
Special District	3.4	3.2	-5.6	0.5	0.7	53.2	7.6	4.8	-37.1	3.8	3.4	-11.3	2.9	2.5	-11.5
All units	30.3	34.5	13.6	21.4	25.2	17.7	34.4	39.5	14.8	19.1	19.5	2.5	26.0	29.4	13.0

Source: U.S. Department of Commerce, Bureau of the Census 2007a.

The number of special districts and county recreation and park departments was much smaller than the number of municipal or township departments, which averaged 3 million residents. (Special districts are authorized by State legislatures to perform a single function or a limited number of functions including but not limited to water and sewage, irrigation, fire control, primary/ secondary/technical education, and hospital administration. Park and recreation services are included among these functions, sometimes as a sole purpose, and at other times as one of many purposes, such as in conservancy.)

Two examples of northern local government agencies that are specifically oriented toward natural resources conservation and recreation are the Wisconsin County Forests and the Illinois County Forest Preserve special districts. The Wisconsin county forests protect more than 2.3 million acres of public forest land in 29 counties and offer a variety of nature-based recreation opportunities. Likewise, in Illinois, the Forest Preserve Districts protect designated lands as "forest preserves" for conservation, education, and "compatible" outdoor recreation experiences. One such example is the Forest Preserve District of Cook County, which manages 68,000 acres of forest, prairie, and wetlands in and around Chicago.

Private Providers

Among the nine outdoor recreation business categories tracked by the Census Bureau's County Business Patterns, five showed a decrease in the number of establishments per million people from 1998 to 2007 (Table 22). In the North, amusement/ theme parks, recreational/vacation camps, golf courses, and marinas all posted decreases, with amusement and theme parks falling by almost half. Private-sector zoos/botanical gardens, nature parks, and historical sites—all in the viewing/learning/photography group of activities—posted the largest gains. Recreational vehicle parks and campgrounds (4.1 percent growth) was the only category that exceeded the national growth rate (1.4 percent).

Current Availability of Nearby Recreation Resources

Federal and State Parkland

Figure 10 shows county-level availability of Federal and State parkland within a 75-mile radius, the distance considered to be the maximum distance of most day trips, with no overnight stay necessary. Whereas some counties have no Federal or State land within their boundaries, all have some public acreage if acreages in surrounding counties are considered. The large majority of counties in the North have fewer than 1,461 acres of public land per 1,000 persons, with numerous counties having less than 70 acres per 1,000 people. These least abundant areas are concentrated around the northeastern metropolises stretching from Maryland to Boston; in western New York; throughout much of Ohio, Indiana, and southern Michigan; in the urban region extending from greater Chicago to Milwaukee; and from the area near metropolitan Kansas City up through western Iowa to the Twin Cities.

Table 22. Number of selected private recreation business establishments per million people in 1998 and percentage change from 1998 to 2007; estimated U.S. population was 272.65 million (Woods and Poole Economics, Inc. 2009) for 1998 and 301.29 million (U.S. Department of Commerce, Bureau of the Census 2009a) for 2007 (includes AK and HI)

Recreation Business entity	North		South		Rocky Mountains		Pacific Coast		United States	
	1998	Percent Change 2007	1998	Percent Change 2007	1998	Percent Change 2007	1998	Percent Change 2007	1998	Percent Change 2007
Golf courses and country clubs	49.8	-4.3	40.1	-11.7	47.1	-6.0	26.0	-8.8	42.6	-7.7
Recreational vehicle parks and campgrounds	13.3	4.1	11.4	0.4	26.9	-1.3	17.3	-3.1	14.5	1.4
Marinas	17.7	-3.5	16.5	-18.6	7.5	-24.9	10.6	-12.3	15.3	-11.5
Recreational and vacation camps (not campgrounds)	14.0	-18.1	8.5	-20.7	22.5	-28.4	11.3	-12.2	12.5	-19.6
Historical sites	4.5	6.1	2.3	23.9	3.8	-15.2	1.8	3.4	3.3	6.7
Nature parks and similar institutions	1.9	31.7	1.5	35.6	2.9	20.7	1.3	118.5	1.7	42.5
Amusement and theme parks	3.7	-47.8	3.6	-36.8	2.9	-16.6	2.4	-12.9	3.4	-37.7
Zoos and botanical gardens	1.4	33.8	1.3	44.8	1.7	15.5	1.5	49.3	1.4	37.8
Skiing facilities	1.7	0.6	0.4	-40.5	4.4	-7.7	1.5	-17.6	1.5	-8.3

Source: U.S. Department of Commerce, Bureau of the Census 2007b.

The relatively more abundant categories of 1,461 to 18,310 acres per 1,000 people are concentrated in a large region extending across northern Minnesota and Wisconsin to the Upper Peninsula of Michigan and the northernmost counties of the Lower Peninsula; smaller pockets also exist in southern Missouri, eastern West Virginia, and northern New Hampshire. Cook County in the Boundary Waters of Minnesota, is the only Northern county in the most abundant category of more than 18,310 acres per 1,000 people.

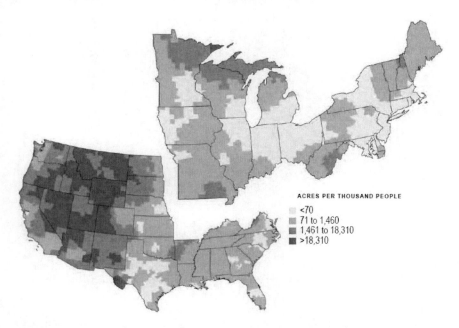

Figure 10. Federal and State parkland area within a 75-mile recreation day trip of each U.S. county 2008 (Sources: U.S. Department of Agriculture Forest Service 2008; U.S. Department of Interior National Park Service 2008; U.S. Department of Interior Bureau of Land Management 2008a; Tennessee Valley Authority 2008; U.S. Army Corps of Engineers 2006; National Association of State Park Directors 2009).

Non-Federal Forests

Residents of the drier counties of western Texas and some parts of Nevada and California lack non-Federal forest land within a 75-mile recreation day trip zone (Fig. 11). Although all northern counties have access to non-Federal forest within the zone, the most abundant non-Federal forests are in northern Minnesota, Wisconsin, Michigan, and much of Maine. In addition to having large forest acreage, these areas are sparsely populated, which means that they have the most non-Federal forest acres per person. With the exception of Missouri, most of the midwestern area is in the category of less than 1,330 acres per 1,000 people. Not surprisingly, the northeastern urban corridor is also in the least abundant category.

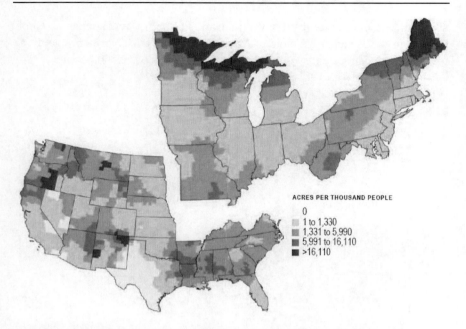

Figure 11. Non-Federal forest area within a 75-mile recreation day trip of each U.S. county, 2010 (Source: U.S. Department of Agriculture Forest Service 2010).

Water

As with public land area, all counties have access to some water area when the 75-mile zone for each county is considered (Fig. 12). Water as defined here is all water area with the exception of open ocean. For the North, greater water area per capita is in the same Great Lakes region of Minnesota, Wisconsin, and Michigan that is abundant in Federal and State parkland and non-Federal forests. The counties of eastern Maine are also among the Nation's most abundant in water acres per capita. Among the least abundant are the Kansas City and St. Louis metropolitan areas, a large portion of central and western central Iowa, and a large band of counties stretching from eastern Illinois all the way to the New York metropolitan area.

Forecasts of Future Availability

Federal and State Parkland

Federal and State parkland area is expected to be constant or almost constant through time. Nearly 30 percent of the total U.S. land and water area is in Federal or State management, which is slightly more than 2 acres per person (Table 23). Because of population growth, per capita Federal and State park acreage is predicted to decrease to 1.4 acres per person (or about 68 percent of the 2008 amount) by 2060.

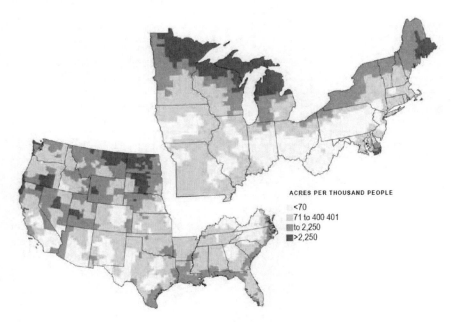

Figure 12. Inland, coastal, territorial, or Great Lakes water area within a 75-mile recreation day trip of each U.S. county, 2008 (Source: U.S. Department of Commerce, Bureau of the Census 2000).

In the North, the Federal or State parkland area per person is projected to decrease to 0.13 acres, about 79 percent of the 2008 level. Compared to their western counterparts, the Eastern States currently have relatively little Federal or State parkland; for example, the North region has only 4.2 percent of national total, which is less than 0.2 acres per person. The percentage of total area that is Federal or State parkland is slightly higher and the population is lower in the north central area than in the northeastern area, but both areas are projected to have the same rate of population growth through 2060 as the North region overall.

Table 23. Projected change in total acres and per capita acres of federal and State parkland with percent of total surface area 2008, projected per capita acres 2060, and percent of 2008 acres projected for 2060, by region (not including AK or HI)

| Region | Federal and state park land[a] | | | | |
| | 2008 | | | 2060 | |
	Acres (thousands)	Percent of total area	Per capita acres	Projected per capita acres	Percent of 2008 acres projected for 2060
North	19,915	4.2	0.16	0.13	79
South	28,274	5.0	0.28	0.17	63
Rocky Mountains	259,643	34.6	9.35	5.22	56
Pacific Coast	319,487	49.5	6.51	4.19	64
U.S. total	627,319	25.8	2.06	1.40	68

[a]Federal and State parkland is the sum of federal land-managing agency area and state park system areas. Federal agencies include NPS, USFS, USFWS, BLM, TVA, and USACE. USDI Bureau of Reclamation not included because most of its areas are managed by other agencies.

Sources: U.S. Department of Agriculture Forest Service 2008, 2009b; U.S. Department of the Interior National Park Service 2008; U.S. Department of the Interior Fish and Wildlife Service 2008; U.S. Department of the Interior Bureau of Land Management 2008; Tennessee Valley Authority 2008; U.S. Army Corps of Engineers 2006.

Non-Federal Forests

Non-Federal forest land area is expected to continue to be converted to developed uses. Excluding Alaska and Hawaii, about 19 percent of total U.S. surface area is non-Federal forest (Table 24), which is about 1.27 acres per person. By 2060, per capita non-Federal forest area is predicted to decrease to 0.8 acres per person, about 63 percent of the current level. In the North, more than 31 percent of total land area is non-Federal forest, which is the highest percent of any region (just slightly ahead of the South). Given the current North population, this represents about 1.19 acres per person. By 2060, per capita non-Federal forest is predicted to decrease to 0.88 acres per person, or 74 percent of the current level.

The percentage of total area that is non-Federal forest is much higher for the northeastern area (53 percent) than for the north central area (22 percent). However, given differences in population growth, projected per capita acreages in each area will be nearly the same by 2060, with both at 0.88 acres per person.

Outdoor Recreation in the Northern United States 49

Table 24. Projected change in total and per capita acres of non-Federal forest land by region for the contiguous United States from 2010 to 2060 (not including AK and HI)

Region	Non-Federal forest land				
	2010			2060	
	Total acres (thousands)	Percent of total area	Per capita acres	Projected per capita acres	Percent of 2010 acres projected for 2060
North	147,762	31.4	1.19	0.88	74
South	171,810	30.5	1.66	0.95	57
Rocky Mountains	28,486	3.8	1.02	0.55	54
Pacific Coast	37,736	17.1	0.79	0.47	59
U.S. total	385,793	19.3	1.27	0.80	63

Source: U.S. Department of Agriculture, Forest Service 2010.

Table 25. Projected change in inland, coastal, territorial, and Great Lakes water area from 2008 to 2060, by region (not including AK or HI)

Region	Water area				
	2008			2060	
	Acres (thousands)	Percent of total area	Per capita acres	Projected per capita acres	Percent of 2008 acres projected for 2060
North	56,834	12.1	0.46	0.36	79
South	29,282	5.2	0.28	0.18	63
Rocky Mountains	7,289	1.0	0.26	0.15	56
Pacific Coast	70,848	11.0	1.44	0.93	64
U.S. total	164,253	6.8	0.54	0.37	68

Source: U.S. Department of Commerce, Bureau of the Census 2000.

Water

Like Federal and State parkland, total water area is expected to be constant or almost constant through time. About 7 percent of total U.S. surface area is water, which roughly equates to a half acre per person (Table 25).

For the North, which is heavily influenced by the Great Lakes, water area is 0.46 acres per person, or more than 12 percent of the total surface area (slightly more than the Pacific Coast and a lot more than the other two regions). By 2060, per capita water is predicted to decrease to 0.36 acres, 79 percent of the 2008 level. Water as a percent of total surface area is slightly higher in the north central (12.9 percent) than in the northeastern area (10.2 percent). Per capita water acreages for the north central area (0.55 acres per capita) are projected to remain much larger than in the northeastern area (0.18 acres per capita), although both areas expect population to grow at the same rate.

SUMMARY OF KEY FINDINGS

Population

- From 1990 to 2009, total population in the North grew at a rate (11 percent) that was less than half the national rate (23 percent). The region has just under 40 percent (14.9 million) of the country's African American population and ranks second to the Pacific Coast in Asian/ Pacific Islander population, which had more than doubled (123 percent) in the North. The Hispanic population almost exactly doubled to about 11.1 million.
- The South and the Rocky Mountains were the only regions to outpace the national growth rate for every single age group. In the North, the Baby Boomer age groups ranging from 45 to 64 have dominated all others in percentage growth from 1990 to 2009, but the rate of growth for this group lagged behind the rest of the Nation. Modest decreases occurred in the North for the younger than 6 age group and held constant for 6- to-10-yearolds. Interestingly, the 25- to 34-age group decreased by nearly 17 percent, which far exceeded decreases in other regions and for the Nation as a whole. This phenomenon was likely a function, in part, of early-career individuals seeking employment opportunities elsewhere. The rate of growth of people age 65 and older in the North (14 percent) was less than half that of the other regions and the Nation, where the growth was 27 percent.
- The North is known for its cluster of high-density metropolitan counties that stretch from greater Washington and Baltimore to southern Maine. Major cities include Philadelphia, New York, and Boston. Other high-density areas include many of the rust belt cities of Pennsylvania and Ohio; most of southern Michigan; the large urban corridor stretching from Gary, IN, through metropolitan Chicago and north to Milwaukee; the Minnesota Twin Cities of Minneapolis and St. Paul; greater Indianapolis; and the two urban regions of Missouri,

St. Louis, and Kansas City. The highest growth in population density (persons per square mile) from 1990 to 2009 occurred in the Washington-to-Boston urban corridor and in the Greater Chicago and Twin Cities areas. Although counties located in suburban areas of the largest cities grew, the urban cores of metropolitan Detroit, Cleveland, Cincinnati, St. Louis, and Buffalo, NY, all lost population. Losses also occurred in much of Pennsylvania, West Virginia, New York, Iowa, and northern Missouri.

- With moderate growth, the total population of the United States is projected to exceed 447 million by 2060, an increase of almost 46 percent. Projected growth for the North is expected to be 26 percent, much less than the other three regions. The 12 States and the District of Columbia in the northeastern area and the eight States in the north central area are all expected to grow at virtually the same rate. New Hampshire, Maryland, and Vermont have the largest projected growth, and the smallest growth rate is expected for New York, West Virginia, and Ohio. Washington, D.C.'s population is projected to decrease almost 17 percent.

Recreation Demand

- One overriding recreation trend is that the relative popularity of outdoor activities is shifting over time. For example, the number of people participating in wildlife or birdwatching and photography is growing while numbers participating in some other activities are not.
- For the North, rate of growth of total outdoor recreation participants and total activity days was lower than the national rate. For those age 16 and older, participation increased about 4 percent, from about 90 to 94 million, and participation days increased by 24 percent in the last decade. Average participation days per person across the full list of 60 activities rose from about 300 per year to 359, a 20 percent increase. Some of the slower gains can be attributed to the North's lower population growth rate compared to the Nation as a whole.
- Of the most popular activities in the North (those having over 30 million participants), the top six slots were occupied by walking for pleasure, attending family gatherings outdoors, gardening or landscaping, viewing/ photographing natural scenery, visiting outdoor nature centers, and picnicking.

Other popular growth activities included viewing/photographing flowers and trees, viewing/photographing wildlife (besides birds and fish), visiting a beach, and viewing/ photographing birds. Activities oriented toward viewing and photographing nature (scenery, flowers/trees/other plants, birds, and wildlife) have been among the fastest growing in popularity.

- Among moderately popular activities (10 to 30 million participants), the most popular were viewing or photographing fish, warm water fishing, motor boating, and visiting a waterside (besides a beach). Growth has been especially strong for off-highway-vehicle driving, warm water fishing, and viewing or photographing fish.
- In the 3 to 10 million participant category, backpacking and ice skating have both been declining over the past 10 years, indicating continuing shifts in activity popularity. Kayaking was the fastest growing of these activities by a wide margin, followed by snowboarding, caving, and water skiing. Several activities posted decreases during this decade.
- Only six activities attracted fewer than 3 million participants, led by orienteering (which grew by over 90 percent in the last decade), snowshoeing, and migratory bird hunting. These are primarily niche activities that appeal to specialized segments of recreation participants.
- Just under three-fourths of northern 6-to19-year-olds spent 2 or more hours outdoors on a typical weekend day (58 percent on a weekday). Forty-three percent spent 4 or more hours outdoors on weekend days (26 percent on weekdays).
- Among young people 6 to 19 years old, unstructured free play or "hanging out" and biking/jogging/walking/skateboarding were the leading outdoor activities, each with more than 78 percent participation. Slightly more than half of respondents used electronic devices while they were outdoors, presumably much of it during unstructured time.
- Among 6- to 19-year-olds, structured nature-based activities, such as attending outdoor camps, classes, and field trips, attracted about 36 percent. Approximately 30 percent also participated in a variety of nature-based recreation activities, such as swimming, diving, snorkeling, birdwatching, wildlife viewing, hiking, camping, and fishing.

Public Recreation Resources

- Compared to the more than 92 percent of Federal land that is located in the West, less than 3 percent, about 17.9 million acres, is in the North and about 69 percent of that is managed by the Forest Service.
- Although Federal acreage changes very little over time, population changes a great deal. In the North, which had an 8 percent decrease, Federal acres per 1,000 persons decreased more slowly than the national rate of decrease (-14 percent since 1995).
- The North accounts for just 1.5 percent of the land area in the National Wilderness Preservation System, about 1.7 million of the over 109 million acres nationally. Modest additions to the system in the North since 1995 resulted in a 1.5 percent increase in wilderness acres per capita.
- Nearly 2,200 miles of National Wild and Scenic River miles are in the North (about 17 percent of all designated river miles), representing a 6.0 percent increase in protected river miles since 2000 (less than the national growth rate of 11 percent).
- The North has more than 7,300 National Recreation Trail System miles, more than any other region and about 36 percent of the system nationally. The addition of 3,200 miles (78 percent) since 2004 was higher than any other region except the South. The region has fewer than 10 Federal recreation facilities per million people, or about 1 per 105,000 people. After camping, boating is the most common Federal recreation facility.
- State park system areas total nearly 5.2 million acres in the North. Throughout the region, especially New England and the rest of the Northeastern States, State park resources are situated within an hour's drive of home for most people.
- Nationwide, more than 8,800 local governments provide recreation and park services. Nearly half (48.3 percent) of these local units were in the North, where the number of local parks and recreation departments per million people was up almost 14 percent since 1997—very close to the national growth rate of 13 percent. Some local government agencies have specific mandates to manage for conservation and compatible nature-based recreation.
- On average, residents of the North have access to fewer than 1,460 acres of public land per 1,000 people (or 1.5 acres per person) within 75 miles of their homes.

- Within a 75-mile recreation day trip zone, the greatest water (non-ocean) area per capita is in counties located near the Great Lakes.
- The North has relatively more non-Federal forest land along the Appalachian Mountains, in southern Illinois, much of Missouri, and similar to water, is most abundant in Maine and northern Minnesota, Wisconsin, and Michigan. On a per capita basis, most of the metropolitan areas have relatively little forest land close by.

Projected Resource Futures

- In the North, the Federal and State parkland area per person is projected to decrease to 0.13 acres, about 79 percent of the 2008 level, by 2060. Because the northern population is not projected to grow as fast as the Nation or any other region, the projected decrease per capita is lower.
- Currently, more than 31 percent of total land area in the North is non-Federal forest, or 1.19 acres per person. By 2060, per capita non-Federal forest is predicted to decrease to 0.88 acres per person, or 74 percent of the 2010 level, lower than all other regions and the Nation as a whole.
- Total water area, like Federal and State parkland, is expected to stay mostly constant over the next several decades. Currently, water area in the North is slightly more than 12 percent of the region's total surface area, or 0.46 acres per person. By 2060, per capita water is predicted to decrease to 0.36 acres per person, or 79 percent of the 2008 level. Similar to the other resources, the projected reduction in water resources per capita is less for the North than for the Nation and all other regions.

DISCUSSION OF FINDINGS

The North has been and continues to be a socially dynamic region of the country. It is a region characterized by large metropolitan areas, population diversity, steady projected population growth, and a mixture of public and private land and water resources. In the last two decades, the North's population grew at a considerably slower rate than the Nation as a whole. Growth has been moderated by many of the region's older people having moved to warmer climates and many of its younger people having moved in search of better employment opportunities.

Even though growth is slower than in other regions, the large population of the North means numerous densely populated communities, large commercial areas, and a wide array of industrial complexes. Many areas have changed radically to accommodate communities and their infrastructure, leaving only a fraction of the natural lands that once dominated the landscape.

At the same time, more individuals, families, and other households translate into greater demand for venues for outdoor recreation. This rising demand presents a dilemma for the North's shrinking supply of undeveloped lands. Will these lands and the developed parks of the future North be sufficient to meet public expectations?

Not only is recreation demand growing, but also what people now choose as outdoor activities is shifting from what they were in past decades and generations. Similar to some of the relatively new activities like orienteering, snowboarding, and mountain biking, which were largely unknown to past generations, new outdoor activities will undoubtedly emerge as the 21st century continues to unfold. One very prominent factor driving this emergence is the changed and ever changing relationship between young people and the outdoors. Contrary to the widely held notion that children in today's United States are not spending time outdoors, the National Kids Survey results suggest that they may actually spend quite a bit of time outdoors, even though significant numbers are using electronic devices when doing so.

Today's youth are tomorrow's adults. How they spend their time now will carry over to affect their future adult lifestyles. Certainly, they have different interests than their parents' generation. The experiences and opportunities of today's young people are different. Likewise, the next generation will be different than the youth of today. If history is a predictor of the future, generational differences will continue to be major drivers of change. This will very likely influence the way people think about and use the outdoors. Undoubtedly, this future use will involve electronic devices, and who can know what these may be in 20 years.

Concurrent with population growth and shifting recreation demands is a very strong likelihood of increasing pressure on forest and other undeveloped lands. Especially in the North, this poses a challenge. Because of high population densities, the average resident of most northern counties has access to fewer than 1.5 acres of Federal or State land within 75 miles of his or her residence. As well, many of the major metropolitan areas have relatively little access to nearby non-Federal forest land, and recreationally accessible water is becoming increasingly scarce throughout much of the region. Like public lands, total water area is fairly static over time; with increasing population, this translates into decreasing per capita acreage in future years.

Population, recreation, and resource trends are all headed in directions that leave one wondering. Who will be the future recreation participants from among the North's growing and changing population? Will participants of the future be representative of the growing diversity of this region's population? Or, could there be a narrowing of participants' demographics as a result of increasing per capita scarcity of places and resources for outdoor recreation?

Where will outdoor recreation occur in the future? As land and water resources in rural areas are increasingly pressured by expanding urban and other development uses, private land and water are likely to become less available for outdoor recreation for some segments of the population. This raises the question of how future residents of the North may gain access to outdoor recreation areas. If the importance of easily accessible, nearby public or private outdoor resources increases in the future, recreation and other nontangible benefits could become important factors in land-value calculations, especially in areas close to population centers. Without inclusion of the value of recreation and other ecosystem services in land value calculations, development value will almost always outweigh other considerations. Including recreation and other ecosystem service values perhaps would open an opportunity for local citizenry and public service organizations to offer incentives that would encourage private owners to keep more land in forest and make it more accessible.

REFERENCES

American Trails. 2010. National recreation trails database. Redding, CA: American Trails. http://www.americantrails.org/NRTDatabase/index.php or http://tutsan.forest.net/trails/. (22 February 2010).

Clawson, M.; Knetsch, J.L. 1966. Economics of outdoor recreation. Baltimore, MD: Resources for the Future. 328 p.

Cordell, H.K. 2008. The latest on trends in nature-based outdoor recreation. Forest History Today. Spring: 4-10.

Cordell, H.K.; Betz, C.J.; Green, G.T. 2008. Nature-based outdoor recreation trends and wilderness. International Journal of Wilderness. 14(2): 7-13.

Cordell, H.K.; Betz, C.J.; Green, G.T.; Mou, S.; Leeworthy, V.R.; Wiley, P.C.; Barry, J.J.; Hellerstein, D. 2004. Outdoor recreation for 21st century America. State College, PA: Venture Publishing, Inc. 293 p.

Cordell, H.K. 2012. Outdoor recreation trends and futures: a technical document supporting the Forest Service 2010 RPA Assessment. Gen. Tech. Rep. SRS-150. Asheville, NC: U.S. Department of Agriculture Forest Service, Southern Research Station. 167 p.

Daigle, J.J.; Utley, L.; Chase, L.C.; Kuentzel, W.F.; Brown, T.L. 2012. Does new large private landownership and their management priorities influence public access in the northern forest? Journal of Forestry. 110(2): 89-96.

Franklin, R.S. 2003. Domestic migration across regions, divisions, and states: 1995 to 2000. Washington, DC: U.S. Department of Commerce, Economics and Statistics Administration, Census Bureau. http://www.census.gov/prod/2003pubs/censr-7.pdf. (17 October 2011).

Interagency Wild and Scenic Rivers Council. 2009. River mileage classifications for components of the national wild and scenic rivers system. Washington, DC: Interagency Wild and Scenic Rivers Coordinating Council. http://www.rivers.gov/publications/rivers-table.pdf. (23 February 2010).

Larson, L.R.; Green, G.T.; Cordell, H.K. 2011. Children's time outdoors: results and implications of the national kids survey. Journal of Park and Recreation Administration. 29(2): 1-20.

National Association of State Park Directors. 1996. Annual information exchange for the period July 1, 1994 through June 30, 1995. On file with: North Carolina State University, Department of Parks, Recreation and Tourism Management, Jordan Hall 5107, Box 7106, Raleigh, NC 27697-7106. (Yu-Fai Leung, principal investigator).

National Association of State Park Directors. 2009. Annual information exchange for the period July 1, 2007 through June 30, 2008. On file with: North Carolina State University, Department of Parks, Recreation and Tourism Management, Jordan Hall 5107, Box 7106, Raleigh, NC 27697-7106. (Yu-Fai Leung, principal investigator).

Outdoor Recreation Resources Review Commission. 1962. National recreation survey. ORRRC Study Report 19. Washington, DC: Outdoor Recreation Resources Review Commission. http://www.srs.fs.usda.gov/trends/Nsre /orrrc.html. (30 June 2011).

Shifley, S.R.; Aguilar, F.X.; Song, N.; Stewart, S.I.; Nowak, D.J.; Gormanson, D.D.; Moser, W.K.; Wormstead, S.; Greenfield, E.J. 2012. Forests of the

Northern United States. Gen. Tech. Rep. NRS-90. Newtown Square, PA: U.S. Department of Agriculture, Forest Service, Northern Research Station. 202 p.

Tennessee Valley Authority. 2008. Recreation resources inventory database. On file with: Tennessee Valley Authority, 400 West Summit Hill Drive, Knoxville, TN 37902.

U.S. Army Corps of Engineers. 2006. Value to the nation: recreation fast facts. http://www.corpsresults.us/recreation/recfastfacts.asp. (1 April 2009).

U.S. Department of Agriculture Forest Service. 1995. Land areas report as of September 30, 1995. On file with: U.S. Department of Agriculture Forest Service, Lands and Realty Management Office, 1400 Independence Ave., SW, Mailstop 1124, Washington, DC 20250-1124.

U.S. Department of Agriculture Forest Service. 2008. Land areas report as of September 30, 2008. Washington, DC: U.S. Department of Agriculture, Forest Service. http://www.fs.fed.us/land/ staff/lar/. (9 February 2009).

U.S. Department of Agriculture Forest Service. 2009a. National survey on recreation and the environment [NSRE Dataset]. On file with: U.S. Department of Agriculture Forest Service, Southern Research Station, Recreation, Wilderness, Urban Forest, and Demographic Trends Research Group, 320 Green St., Athens, GA 30602-2044.

U.S. Department of Agriculture Forest Service. 2009b. State park systems database compiled from published state literature and state park Web sites. On file with: U.S. Department of Agriculture Forest Service, Southern Research Station, Recreation, Wilderness, Urban Forest, and Demographic Trends Research Group, 320 Green St., Athens, GA 30602-2044.

U.S. Department of Agriculture Forest Service. 2010. RPA Assessment land use projections. On file with: U.S. Department of Agriculture Forest Service, Southern Research Station, Forest Economics and Policy, P.O. Box 12254, Research Triangle Park, NC 27709.

U.S. Department of Commerce, Bureau of the Census. 1990. Census of population and housing. 1990. Summary tape file 1. http://www2.census.gov/census_1990/1990STF1.html#1A. Washington, DC: U.S. Department of Commerce, Bureau of the Census. (23 September 2009).

U.S. Department of Commerce, Census Bureau. 2000. Census 2000 U.S. gazetteer files, water area (square miles). Washington, DC: U.S. Department of Commerce, Bureau of Census. http://www.census.gov/geo /www/gazetteer/places2k.html. (20 June 2005).

U.S. Department of Commerce, Census Bureau. 2007a. Census of governments, government employment and payroll, local government, 1997 and 2007. Washington, DC: U.S. Department of Commerce, Census Bureau. http://www.census.gov/govs/apes/. (31 March 2009).

U.S. Department of Commerce, Census Bureau. 2007b. Economic census, county business patterns, 1998 and 2007. Washington, DC: U.S. Department of Commerce, Bureau of Census. http://www.census.gov/econ /cbp/index.html. (24 August 2009).

U.S. Department of Commerce, Census Bureau. 2009a. SC-EST2009-alldata6: annual state resident population estimates for 6 race groups (5 race alone groups and one group with two or more race groups) by age, sex, and Hispanic origin: April 1, 2000 to July 1, 2009. Washington, DC: U.S. Department of Commerce, Bureau of Census. http://www.census.gov/ compendia/statab/2011/tables/11s0019.xls . (9 January 2012).

U.S. Department of Commerce, Census Bureau. 2009b. CC-EST2009-ALLDATA-[ST-FIPS]: annual county resident population estimates by age, sex, race, and Hispanic origin: April 1, 2000 to July 1, 2009. Washington, DC: U.S. Department of Commerce, Bureau of Census. http://www.census.gov/popest/data/counties/asrh/2009/CC-EST2009-alldata.html . (9 January 2012).

U.S. Department of the Interior. 2009. Recreation one-stop initiative: recreation information data base. Washington, DC: U.S. Department of Interior. https://www.recdata.gov/RIDBWeb/ Controller.jpf. (3 April).

U.S. Department of the Interior, Bureau of Land Management. 1994. Public land statistics 1993. Vol. 178. Washington, DC: U.S. Department of Interior.

U.S. Department of the Interior, Bureau of Land Management. 2008. Public land statistics 2008. Washington, DC: U.S. Department of Interior, Bureau of Land Management. http://www.blm.gov/public_land_statistics/pls08 /index.htm. (23 June 2009).

U.S. Department of the Interior, Bureau of Reclamation. 1993. Recreation fast facts. On file with: U.S. Department of the Interior, Bureau of Reclamation, Land & Mineral Records System, 1849 C St., NW, Rm. 5625, Washington, DC 20240.

U.S. Department of the Interior, Bureau of Reclamation. 2008. Recreation fast facts. Washington, DC: U.S. Department of Interior, Bureau of Reclamation. http://www.usbr.gov/recreation. (19 February 2009).

U.S. Department of the Interior, Fish and Wildlife Service. 1995. Annual report of lands as of September 30, 1995. On file with: U.S. Department of the Interior, Fish and Wildlife Service, 4401 N. Fairfax Dr., Arlington, VA 22203.

U.S. Department of the Interior, Fish and Wildlife Service. 2008. Annual report of lands as of September 30, 2008. Washington, DC: U.S. Department of Interior, Fish and Wildlife Service. http://www.fws.gov/refuges/land/LandReport.html. (18 June 2009).

U.S. Department of the Interior, National Park Service. 1995. Listing of acreage by state and county as of 10/31/95. On file with: U.S. Department of the Interior, National Park Service, Land Resources Division, 1849 C St., NW, Washington, DC 20240.

U.S. Department of the Interior, National Park Service. 2008. Listing of acreage by state and county as of 12/31/2008. http://www.nature.nps.gov/stats/acreagemenu.cfm. (26 February 2009).

Wilderness.net. 2009. Wilderness data search. Missoula, MT: Wilderness Institute. http://www.wilderness.net/index.cfm?fuse=NWPS&sec=adv Search. (6 July).

Woods & Poole Economics, Inc. 2009. 2010 complete economic and demographic data source (CEDDS). Washington, DC: Woods & Poole. [CD-ROM].

Yang, T.; Snyder, A.R. 2007. Population change in the Northeast, 2000-2005. State College, PA: University of Pennsylvania. http://nercrd.psu.edu/Publications/rdppapers/rdp39.pdf. (17 October 2011).

Zarnoch, S.J.; Cordell, H.K.; Betz, C.J.; Bergstrom, J.C. 2010. Multiple imputation: an application to income nonresponse in the national survey on recreation and the environment. Res. Pap. SRS–49. Asheville, NC: U.S. Department of Agriculture Forest Service, Southern Research Station. 15 p.

APPENDIX: METHODS AND DATA SOURCES

Population and Demographic Trends and Projected Futures

Historical data from the 1990 U.S. Census of Population and Housing through the 2009 census population estimates were analyzed to examine recent trends in population, population distribution, and demographic composition. National and regional population totals and percents are presented in tables, along with maps showing the distribution of the population among northern counties. All maps show four shading levels that correspond to the following

Outdoor Recreation in the Northern United States 61

percentage distributions of the data depicted in each map: 0 to 35, 36 to 70, 71 to 90, and 91 to 100. The two highest percentage ranges (shown by the darkest shades) are purposely more restricted to emphasize counties having the largest counts of population or more significant data values.

Included in this report are data on population by race/ethnicity, population by age groups, current population density (persons per square mile), change in population density since 1990, percentage change in Hispanic population, percentage change in non-Hispanic White population, and projected changes in population density from 2008 to 2060. For comparison with the North, the same statistics are also shown for all counties of the rest of the country, except for Hawaii and Alaska. The northern region consists of Connecticut, Delaware, District of Columbia, Illinois, Indiana, Iowa, Maine, Maryland, Massachusetts, Michigan, Minnesota, Missouri, New Hampshire, New Jersey, New York, Ohio, Pennsylvania, Rhode Island, Vermont, West Virginia and Wisconsin.

The U.S. Department of Commerce, Bureau of the Census provides updated annual population estimates for the Nation, States, and counties each year. Based on these updates, county-scale maps were produced for this report showing change in Hispanic and other segments of the North's population. (The census released preliminary 2010 estimates of total population by county in March 2011, but had not released population estimates by demographic categories at the time of writing.) Data consulted included:

U.S. Department of Commerce, Bureau of the Census (2009a), SC-EST2009-alldata6: Annual State Resident Population Estimates for 6 Race Groups (5 Race Alone Groups and One Group with Two or more Race Groups) by Age, Sex, and Hispanic Origin: April 1, 2000 to July 1, 2009 (http://www.census.gov/compendia/statab/2011/ tables/11s0019.xls)

U.S. Department of Commerce, Bureau of the Census (2009b), CC-EST2009-ALLDATA-[ST-FIPS]: Annual County Resident Population Estimates by Age, Sex, Race, and Hispanic Origin: April 1, 2000 to July 1, 2009 (http://www.census.gov/ popest/data/counties/asrh/2009/CC-EST2009-alldata.html)

State and county population from the 1990 census were derived from Woods & Poole Economics, Inc. (2009).

Working from Census Bureau estimates, the U.S. Department of Agriculture-Forest Service Southern Research Station developed county-scale forecasts of population change for three of the future scenarios defined by the International Panel on Climate Change in its Fourth Assessment Report (Zarnoch et al. 2010). The scenarios—labeled A1B, A2, and B2—were adapted for use in both the national 2010 Renewable Resources Planning Act (RPA) Assessment (Cordell 2012) and for the Northern Forest Futures Project, currently underway at the Forest Service (Northern Research Station, Eastern Region, Forest Products Laboratory, and Northeastern Area State and Private Forestry) in partnership with the Northeastern Area Association of State Foresters and the University of Missouri. The overall purpose for examining population change in the context of these scenarios is to evaluate the sensitivity of forest and other resource trends to a range of feasible futures. In this report, percentage change over the 50-year assessment period (2010 to 2060) is shown only for the A1B moderate population growth scenario. Under this scenario, total population in the United States is projected to exceed 447 million people by 2060, a growth of almost 46 percent.

Recreation Activity Trend Data

The source of data on recreation activity trends for adults is the National Survey on Recreation and the Environment (NSRE). Sponsored by the Southern Research Station, the University of Georgia and the University of Tennessee, it is a general population random-digit-dialed telephone survey that asks Americans age 16 and older about their participation in outdoor recreation activities (U.S. Department of Agriculture Forest Service 2009a). The NSRE data presented here are from surveys conducted continuously from 1999 to 2009, with a brief interruption during 2004.

Earlier estimates of trends in outdoor recreation in general and in nature-based outdoor recreation in particular (Cordell 2008) were conducted for the RPA Assessment. This report updates those findings. NSRE data were pooled to define two trend periods: 1999 to 2001 and 2005 to 2009. (The volume of NSRE surveying decreased in the latter years which resulted in smaller yearly sample sizes and thus the combining of more years in the later period.) An overview of Americans' participation in outdoor recreation in general was constructed by defining a "participant" as any person who engaged in at least one of 60 outdoor recreation activities one or more times during the 12 months prior to the date they were interviewed. A "yes" value was assigned to respondents if they reported participation in any of the 60 activities, with "no" indicating that the individual did not participate in any activity during the past year. A similar indicator was used to determine nature-based activity participation using a shorter list of 50 activities that typically occur in natural settings. Previous estimates from the 1994-to-1995 period are included to indicate overall trends across two decades.

The source of data for youth time in the outdoors is the National Kids Survey (Cordell 2012), a household telephone survey that was conducted by the Southern Research Station in cooperation with the University of Tennessee and the University of Georgia from 2007 to 2011. Households with a 6- to 19-year-old qualified to participate. If a household had more than one qualifying household member, the survey questions were directed to the individual with the most recent birthday. Teenagers 16 to 19 responded for themselves; an adult proxy, usually a parent, answered for children age 6 to 15. The survey data were post-weighted to approximate census population percents by gender and by eight age strata. Questions asked included the amount of time spent outdoors regardless of activity, as well as the types of activities engaged in. A total of 1,945 respondents or their proxies participated, 763 from Northern States.

Recreation Resource Data

Federal Resources
The Federal land managing agencies are the sources for Federal outdoor recreation resources data. The four largest Federal land-managing agencies—Forest Service and the U.S. Department of the Interior Bureau of Land Management, Fish and Wildlife Service, and National Park Service—have real estate offices that maintain records on the size, location, and boundaries of their holdings. The three Federal water resources management agencies—U.S. Army Corps of Engineers, U.S. Department of the Interior Bureau of Reclamation, and Tennessee Valley Authority— have much smaller land holdings.

Resources protected by inclusion in the National Wilderness Preservation System, the National Wild and Scenic River System, and the National Trails System are also described in this paper in the section titled specially designated Federal land systems. Current and past data from each of these systems was examined for trends in per capita availability.

Federal recreation sites and facilities are cataloged in an online database called the Recreation Information Database, better known through its portal as www.recreation.gov [Date accessed: December 15, 2011] or simply rec. gov. The Department of the Interior coordinates an interagency coalition that gathers recreation site and facilities information across all Federal agencies.

64 H. Ken Cordell, Carter J. Betz, Shela H. Mou et al.

The rec.gov Web site includes a standardized list of 22 separate recreation activities, facilities, or attractions with binary (yes/no) availability. Trend data are not available for this database because it is fairly new, originating around 2002. Both a limitation and a strength of the Recreation Information Database is that it is an evolving source of information, which is expanding and growing, but not yet complete across Federal recreation facilities.

State Resources

Two sources of State park system data were used in this report. First is the National Association of State Park Directors Annual Information Exchange survey, which collects data from all 50 State park systems. This report uses survey data to assess the status of each State park system's resources, operations, and visits. Included in coverage are State parks, recreation areas, natural areas, historical areas, environmental education areas, scientific areas, forests, wildlife and fish areas, and other miscellaneous areas. The exchange summarizes all information by State; it does not have individual State park unit information, such as size, location, and site attributes. The most consistent data over the history of the exchange have been about the State park and State recreation area classifications.

The second source is a State park database of individual park system units developed from printed and online sources[†], and includes acreage data and location (latitude/longitude coordinates). The database focuses on the three most common types of State park system units: parks, recreation areas, and historic sites.

Local Government Resources

The data source for local government outdoor recreation resources is the Census Bureau Census of Governments, which is conducted every 5 years. The classifications for this census are type of governmental unit and services provided. A difficulty in assessing local government recreation resources is the sheer number and variety of local jurisdictions that provide park and recreation services. Further, many local agencies place as much (and sometimes more) emphasis on indoor leisure programs and services as on outdoor resources. The Census of Governments does not provide details on land holdings or other resources; rather it collects administrative, financial, and employment data. This report assumes that all local government agencies listed as providing

---

[†] Cordell, H. Ken. 2011. [Untitled]. Unpublished database. On file with: Pioneering Research Project, Southern Research Station, U.S. Department of Agriculture Forest Service, 320 Green Street, Athens, GA 30602-2044.

recreation and park services also include management of some outdoor recreation resources, although the amount provided is not known.

Private Recreation Businesses
The Census Bureau provides the annual County Business Patterns (CBP) series of economic data. Included is number of recreation business establishments, payroll, and number of employees for the full range of businesses as described in the North American Industry Classification System. Nine of the business classes listed are related to outdoor recreation. Summarized in this report are number of business establishments per capita, along with percentage change from the 1998 to the 2007.

County Pattern Maps
Included in this report are county-level maps for 2008 that depict patterns of recreation resource availability per capita across northern counties and the rest of the Nation. (These recreation resource maps employ the same criteria as was used with the demographic data, which displays four shading levels based on percentage distributions of the data: 0 to 35, 36 to 70, 71 to 90, and 91 to 100.) Recreation resources per capita within a 75-mile radius of each county are displayed in the maps.

The 75-mile zone includes a home county plus all surrounding counties whose geographic centers or centroids are within a 75-mile straight-line distance from the home county centroid. This distance is roughly the equivalent of a recreation day trip. The three basic recreation resources summarized in this report are combined Federal land and State park area, non-Federal forest land, and water area (from census Tiger geographic data).

Projected Futures

Using the population projections described earlier, projections of per capita recreation resources were developed for three resources. The projection index used is the ratio of per capita acres predicted for 2060 relative to the per capita acres that existed in base year 2008. This statistic indicates the percent of the resource currently available per capita that is forecast to remain by 2060. The per capita future is forecast for three recreation resources—Federal and State parkland, non-Federal forest land, and water. Projections are summarized for the North and for other regions. Also reported is the percentage of total surface area in each region represented by the resource.

REFERENCES

Cordell, H.K. 2008. The latest on trends in nature-based outdoor recreation. Forest History Today. Spring: 4-10.

U.S. Department of Agriculture Forest Service. 2009a. National survey on recreation and the environment [NSRE Dataset]. On file with: U.S. Department of Agriculture Forest Service, Southern Research Station, Recreation, Wilderness, Urban Forest, and Demographic Trends Research Group, 320 Green St., Athens, GA 30602-2044.

U.S. Department of Commerce, Census Bureau. 2009a. SC-EST2009-alldata6: annual state resident population estimates for 6 race groups (5 race alone groups and one group with two or more race groups) by age, sex, and Hispanic origin: April 1, 2000 to July 1, 2009. http://www.census.gov /compendia/statab/2011/tables/11s0019.xls. (9 January 2012).

U.S. Department of Commerce, Census Bureau. 2009b. CC-EST2009-ALLDATA-[ST-FIPS]: annual county resident population estimates by age, sex, race, and Hispanic origin: April 1, 2000 to July 1, 2009. http://www.census.gov/popest/data/counties/asrh/2009/CC-EST2009-alldata.html. (9 January 2012).

Woods & Poole Economics, Inc. 2009. 2010 complete economic and demographic data source (CEDDS) .Washington, DC. Woods & Poole Economics, Inc. [CD-ROM].

Zarnoch, S.J.; Cordell, H.K.; Betz, C.J.; Bergstrom, J.C. 2010. Multiple imputation: an application to income nonresponse in the national survey on recreation and the environment. Res. Pap. SRS–49. Asheville, NC: U.S. Department of Agriculture Forest Service, Southern Research Station. 15 p.

In: Outdoor Recreation ...
Editor: Alexander N. Borden

ISBN: 978-1-63117-110-9
© 2014 Nova Science Publishers, Inc.

Chapter 2

OUTLOOK FOR OUTDOOR RECREATION IN THE NORTHERN UNITED STATES. A TECHNICAL DOCUMENT SUPPORTING THE NORTHERN FOREST FUTURES PROJECT WITH PROJECTIONS THROUGH 2060[*]

J. M. Bowker and Ashley E. Askew

ABSTRACT

We develop projections of participation and use for 17 nature-based outdoor recreation activities through 2060 for the Northern United States. Similar to the 2010 Resources Planning Act (RPA) assessment, this report develops recreation projections under futures wherein population growth, socioeconomic conditions, land use changes, and climate are allowed to change over time.

Findings indicate that outdoor recreation will likely remain a key part of the region's future social and economic fabric. The number of participants in 14 of the 17 recreation activities is projected to increase over the next five decades. In about two-thirds of 17 activities, the participation rate will likely decrease, but population growth would ensure increases in the number of adult participants. Some climate futures

---

[*] This is an edited, reformatted and augmented version of The United States Department of Agriculture, Forest Service, Northern Research Station publication, dated September 2013.

could lead to participant decreases for certain activities. Hunting, snowmobiling, and undeveloped skiing appear to be the only activities for which a decrease in participants is likely. Total days of participation would generally follow the pattern of participant numbers. With the exceptions of hunting, visiting primitive areas, and whitewater activities, snowmobiling, undeveloped skiing, total days are expected to increase for the remaining 14 activities, some less so than others because of climate differences.

INTRODUCTION

This report addresses a major question considered relevant to the Northern Forests Futures Project, namely, "How will population growth, along with changing socioeconomic conditions, demographics, land uses, and climate, influence associated demand for natural resource-based recreation?" The question is addressed through an analysis of natural resource-based outdoor recreation demand for the 20 states that make up the U.S. North (Fig. 1), a region that is bounded by Maine, Maryland, Missouri, and Minnesota.

Figure 1. Regions for the 2010 U.S. Resources Planning Act (RPA) assessment.

The report mirrors and extends previous studies that were part of the recent U.S. Forest Service 2010 Resources Planning Act (RPA) assessment (Bowker et al. 2012) and the Southern Forest Futures Project (Bowker et al., in

Outlook for Outdoor Recreation in the Northern United States 69

press). Specifically, we developed regional projections of participation and use for 17 natural resource-based outdoor-recreation activities, or activity composites (Table 1), through 2060. The report was also designed to complement a concurrent study of current and recent trends in outdoor recreation in the North (Cordell et al. 2012).

Participation and Use

We defined a participant in an outdoor recreation activity as an adult resident over the age of 16 who engaged in that activity at least once in the previous 12 months. Participation is a general indicator of the size of a given recreation market, and it also can be a gauge of public interest. Land managers and legislators can benefit from knowing how many people participate in a given recreation activity as well as how participation could change over time and affect both public support and potential ecological and social carrying capacities (Dale and Weaver 1974, Manning 1997). For example, if more than 80 percent of the population participates in hiking but just 4 percent participate in snowmobiling, public resource management agencies and private land managers may see a greater need for hiking trails than for snowmobiling trails. Measures of participation, either per capita (participation rates) or in absolute numbers of participants, provide the broadest measure of a recreation market.

Table 1. Outdoor recreation activities for 2008 by participants, participation rate, days, and days per participant for Northern residents

Activity	Participants (millions)[b]	Percent Participating	days (millions)[c]	days per Participant
developed site use				
Visiting developed sites – family gatherings, picnicking, developed camping	80.5	82.5	943	11.7
Visiting Interpretive sites – nature centers, zoos, historic sites, prehistoric sites	67.0	68.6	516	7.7

Table 1. (Continued)

Activity	Participants (millions)[b]	Percent Participating	days (millions)[c]	days per Participant
observing nature				
Birding – viewing and/or photographing birds Viewing[d] – viewing, photography, study, or nature gathering related to fauna, flora, or natural settings	37.2 79.5	38.2 81.5	3,696 13,925	99.8 175.7
Backcountry Activities				
Challenge Activities – caving, mountain biking, mountain climbing, rock climbing	9.4	9.5	37.7	3.9
equestrian	5.8	5.9	72.3	12.6
Hiking – day hiking	32.4	32.7	723.8	22.4
Visiting Primitive Areas – backpacking, primitive camping, wilderness	36.1	36.7	415	11.4
Motorized Activities				
Motorized off-road use	17.3	17.6	282.8	16.4
Motorized snow use (snowmobiling)	7.0	7.1	54.8	7.9
Motorized water use	26.1	26.8	378.8	14.7
Hunting and fishing				
Hunting – small game, big game, migratory bird, other	11.3	11.7	209.6	18.8
Fishing – anadramous, coldwater, saltwater, warmwater	28.7	29.6	515.7	18.1
non-Motorized Winter Activities				
downhill skiing – downhill skiing, snowboarding	11.6	11.6	81.3	7.0
undeveloped skiing – cross-country skiing, snowshoeing	4.8	4.8	32.1	6.7

Activity	Participants (millions)[b]	Percent Participating	days (millions)[c]	days per Participant
non-Motorized Water Activities				
swimming – swimming, snorkeling, surfing, diving, visiting beaches or watersides	61.7	63.3	1,376	22.2
Floating – canoeing, kayaking, rafting, sailing	18.2	18.7	124	6.8

Source: NSRE 2005-2009, Versions 1 to 4 (January 2005 to April 2009), n=24,073.
[a] Activities are individual or activity composites derived from the NSRE.
[b] Participants are determined by the product of the average weighted frequency of participation by activity for NSRE data from 2005-2009 and the adult (>16) population in the US during 2008 (235.4 million).
[c] Days are determined by the product of the weighted conditional average days per adult participant and the number of participants by activity for NSRE data from 2005-2009.
[d] Including birding.

A second measure of recreation use is consumption or participation intensity. Consumption can be measured in number of times, days, visits, or trips within a time span, such as a 12-month period. The U.S. Forest Service has used such consumption measures as recreation visitor days and national forest visits. Consumption measures of participation (knowing how often and for how long people engage in an activity) provide an important additional dimension for resource managers who need to know how best to allocate resources, such as campsites, and whether to plan new ones.

Participation and consumption at the regional level provide the broadest measures of an outdoor recreation market. The consumption measure used in this study is the number of days in the previous year that an adult resident of the Northern United States reported engaging in a specific activity. A *day*, in this study, follows the National Survey on Recreation and the Environment (NSRE) definition of an activity day—any amount of time spent on an activity on a given day, whether or not that activity was the primary reason for the outdoors visit. Hence, camping at an improved facility for one night would constitute two days of camping. A person may engage in more than one activity per day, and thus, a person's activity day total per year may not exceed 365 for any specific activity but it may do so when all activities are combined (Cordell 2012).

These two metrics—participation and consumption—are origin based, meaning that they result from household-level surveying. There is no additional information on exactly where the respondent engaged in the participation for any activity, although research shows that the vast majority of outdoor recreation takes place within a few hours' drive of home (Hall and Page 1999). Participation rates and participant numbers for 2008, along with total days spent participating and average days per participant, for the 17 outdoor recreation activities examined in this study are displayed in Table 1.

A history of outdoor recreation trends is an important indicator of what may happen in the near future (Cordell 2012, Hall et al. 2009). However, simple descriptive statistics or trends do not formally address underlying factors and associations that may be driving these trends. Thus, a trend may be of limited value as an indicator if the time horizon is long or if the driving factors are expected to deviate substantially from historical levels. Trend analysis, therefore, can be supplemented with projection models that relate participation directly to factors known to influence participation behavior. The projection models then can be used in conjunction with external projections of relevant factors, including population growth, to simulate future recreation participation and consumption.

Such modeling allows changes in recreation participation and consumption behavior to be assessed in light of large changes in demographics, economic conditions, land use, climate, and other previously unseen influential factors.

Previous research has demonstrated that race, ethnicity, gender, age, income, supply, and proximity to settings may be related to the rate of outdoor recreation participation as well as the participation intensity or consumption (Bowker et al. 1999, Bowker et al. 2006, Cicchetti 1973, Hof and Kaiser 1983b, Leeworthy et al. 2005). Along with distance and quality descriptors and other factors, these have been used to explain visits to specific sites (Bowker et al. 2007, Bowker et al. 2010, Englin and Shonkwiler 1995, Ovaskainen et al. 2001). Reliable information about these factors is often available from external sources such as the U.S. Census Bureau or from parallel research efforts to model and simulate influential variables into the future. Such information thus can be available long before recreation survey results are published.

A two-step approach was used to project participation and consumption of 17 traditional outdoor recreation activities (Table 1). The first step, model estimation, focused on developing regional level statistical models of adult per capita participation and days of participation (conditional on being a participant) for each activity, with the participation model describing the probability of an individual participating in a specified activity and the consumption model describing the number of days of participation for those activities in which an individual participated. This information improves understanding of the influences on individual recreation choices or behaviors and of the way that individual recreation choices or behavior might respond to changes in underlying factors such as demographics, resource availability, and climate.

The second step, or simulation step, combined the estimated models with external projections of relevant explanatory variables to generate estimated per capita participation probabilities and conditional expected days of participation for each activity at 10-year intervals to 2060. These were combined with population projections to develop regional estimates for each activity, which were used to create indices by which 2008 baseline estimates of participants and days of participation for the various activities (Table 1) could be scaled.

The resulting indices of estimated adult participants for each of the 17 activities and days of annual participation are presented for an updated version of an emissions storyline (high economic growth with moderate population growth) from the Intergovernmental Panel on Climate Change in combination with three associated climate futures that were derived from the three general circulation models— CGCM3.1 and CSIROMK3.5 downloaded from the World Climate Research Program Climate Model Intercomparison Project 3 website, and MIROC3.2 downloaded from the IPCC Data Distribution Centre (Joyce et al., in press)— described below.

Model Development

The conditions during the projection period for this report are based on one of three scenarios used for the 2010 RPA assessment (U.S. Department of Agriculture Forest Service 2012). Three RPA scenarios were developed to describe alternative national and county-level futures which were linked to emissions storylines developed during the Intergovernmental Panel on Climate Change third and fourth assessments (IPCC 2007), thereby providing context and quantitative linkages between national and global trends including assumptions and projections of global population growth, economic growth, bioenergy use, and climate (Alcamo et al. 2003, IPCC 2007, Nakic'enovic' et al. 2000).

Table 2. Key characteristics of the Resources Planning Act (RPA) scenarios

Characteristic[a]	scenario RPA AIB	Scenario RPA A2	scenario RPA B2
General global description	Globalization, economic convergence	regionalism, less trade	slow change, localized solutions
Global real GdP[b] growth (2010-2060)	High (6.2X)[c]	Low (3.2X)	Medium (3.5X)
Global population growth (2010-2060)	Medium (1.3X)	High (1.7X)	Medium (1.4X)
IPCC global expansion of primary biomass energy production	High	Medium	Medium
u.s. GdP growth (2006-2060)	High (3.3X)	Medium (2.6X)	Low (2.2X)
u.s. population growth (2006-2060)	Medium (1.5X)	High (1.7X)	Low (1.3X)

[a] Global characteristics are based on Intergovernmental Panel on Climate Change (2007) emissions assumptions. U.S. characteristics are from the 2010 Resources Planning Act (RPA) assessment.

[b] GDP = Gross Domestic Product.

[c] Numbers in parentheses are the factors of change in the projection period. For example, world GDP (gross domestic product) increases by a factor of 6.2 times between 2010 and 2060 for scenario RPA A1B.

Of the three storylines—A1B (high economic growth, moderate population growth), A2 (moderate economic growth, high population growth), and B2 (low economic growth, low population growth)—only A1B was used for this study. Table 2 summarizes key global and national characteristics of all three storylines.

Population projections were developed for each RPA scenario. Projections for the original A1B storyline were based on the 1990 Census.

These were updated to align with the 2004 Census population series for 2000 to 2050 (U.S. Census Bureau 2004), with an extrapolation to 2060. The population projections for the original A2 and B2 were updated to begin at the same starting point, in year 2000, and to then follow a projection path that maintains the same proportional relationship to A1B as in the original projections. Figure 2 illustrates the population projections for the three updated storylines relative to historical population trends (Zarnoch et al. 2010).

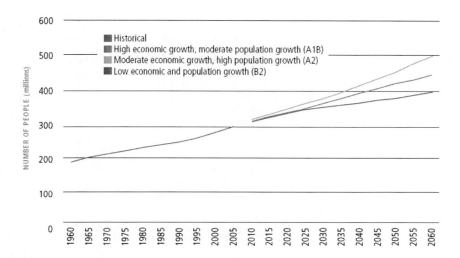

Figure 2. Historical U.S. population and projections through 2060 from the 2010 Resources Planning Act assessment.

Macroeconomic trends—such as Gross Domestic Product (GDP), disposable personal income, and labor productivity—critically influence the supply and demand of renewable resources, and thus, also influence recreation demand. Because the original storylines were based on economic data from the early 1990s, GDP projections were updated to start with the official GDP value for 2006 (U.S. Department of Commerce 2008a). GDP growth rates, provided by a commissioned report, were applied to develop an adjusted projection for A1B. Revised A2 and B2 projections maintained the same proportional relationship among the three storylines as defined by the original GDP projections (U.S. Department of Agriculture Forest Service 2012). Figure 3 shows the differences among the three projections for updated GDP in comparison to historical records.

Projections of personal income and disposable personal income also were developed. U.S. 2006 personal income and disposable personal income data were used to start the updated projection for A1B (U.S. Department of Commerce 2008b). A2 and B2 projections for personal income and disposable personal income maintained the same proportional relationship that was used to calculate the trajectories for GDP. The national disposable personal income and personal income projections were then disaggregated to the county level (U.S. Department of Agriculture Forest Service 2012).

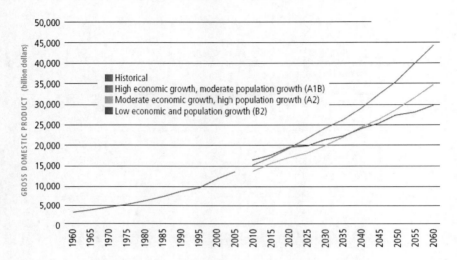

Figure 3. Historical U.S. gross domestic product (GDP) and projections through 2060 from the 2010 Resources Planning Act assessment (2006 U.S. dollars).

The RPA projections were completed before the global economic downturn that began in 2008. Because data from 2006 were the most recent, that year was chosen as the base year for economic variables. The projection trend line from 2006 to 2010 does not account for the downturn in GDP and other economic variables through 2010, creating a discontinuity in the early years of the projection period. Long-term projections are not intended to predict economic ups and downs, meaning that periodic economic recessions would not be a part of the projected 50-year trend. Although the recent global recession was severe, the range of alternatives included in the RPA assessment have varying rates of economic growth, both for the United States and globally, providing a robust set of projections across the range of potential futures.

Land use change is a key factor in outdoor recreation participation and demand. Land use change was projected for all counties in the contiguous United States in five major land use classes: pasture, cropland, forest land, rangeland, and urban and developed uses (Wear 2011). Within these categories, no land use change was assumed to occur on Federal land; additionally, uses were held constant over the projection period for water area, enrolled Conservation Reserve Program lands, and utility corridors for fuels, water, and electricity.

The projected changes in major land uses at the national level for A1B are summarized in Figure 4. This pattern of change is similar for A2 and B2, but with smaller changes than A1B (U.S. Department of Agriculture Forest Service 2012).

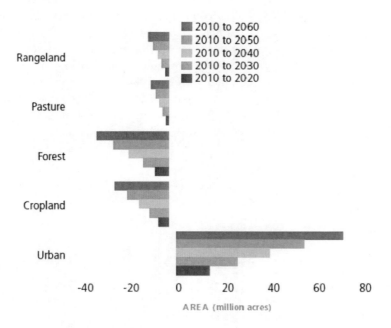

Figure 4. Projected U.S. land use change by major categories, 2010 to 2060, under a future of high economic growth and moderate population growth (A1B); no land-use change was assumed to occur on Federal land, and uses were held constant for water area, enrolled Conservation Reserve Program lands, and utility corridors for fuels, water, and electricity (source: 2010 Resources Planning Act assessment).

In all, increases in urban and developed uses are expected to be the dominant force in land use change, with other land uses projected to lose area accordingly. The highest rate of urbanization is associated with A1B and the lowest with B2, suggesting that strong growth in personal income combined with moderate population growth creates more development pressure than population growth alone. Urban and developed area would increase by 69 million acres from 2010 to 2060 for A1B, almost doubling the amount of urban area over the projection period (Wear 2011).

Forest land would decrease by almost 31 million acres over the projection period under A1B, compared to 16 million acres under B2 (Wear 2011). The South (Fig. 5) is projected to experience the largest decrease in forest area by 2060, about 17 million acres in A1B.

These large losses reflect both a history of comparatively abundant forest resources and a future likelihood of comparatively high population growth and urbanization. The North has the second largest loss of forest land in A1B (almost 10 million acres), followed by smaller losses in the Rocky Mountains and Pacific Coast. Although losses of forest land are smaller in A2 and B2, the pattern of forest land loss is similar for all regions; the exception being the Pacific Coast where A2 predicts higher forest loss than A1B, but the difference is quite small (Wear 2011). Moreover, public forest and rangeland are expected to remain relatively static over the projection period.

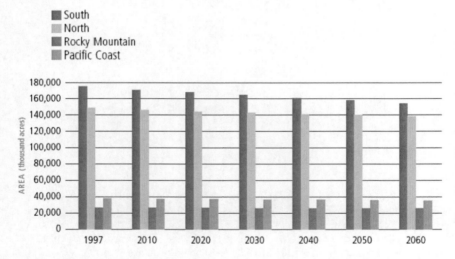

Figure 5. Projected change in U.S. forest land by region, 2010 to 2060, under a future of high economic growth and moderate population growth (A1B); no land-use change was assumed to occur on Federal land, and uses were held constant for water area, enrolled Conservation Reserve Program lands, and utility corridors for fuels, water, and electricity (source: 2010 Resources Planning Act assessment).

After private forest land, cropland is expected to lose the most acreage, mostly in the Eastern United States, which currently has more cropland than the Western States. Cropland losses are nearly equally split between the North and South (Wear 2011). Rangeland losses are concentrated in the Rocky Mountains, which has about half the total rangeland losses. The remainder of rangeland losses is split between the South (primarily in Texas) and the Pacific Coast (mostly southern California).

Few large-scale studies have been conducted to relate climate to outdoor recreation, but an underlying assumption of this report is that long-term climate changes can affect recreation demand. Each storyline from the Intergovernmental Panel on Climate Change had multiple associated climate projections, which varied in response to the associated levels of greenhouse gas emissions. They also varied because of differences in the general circulation models that were associated with them in the RPA assessment (Joyce et al., in press).

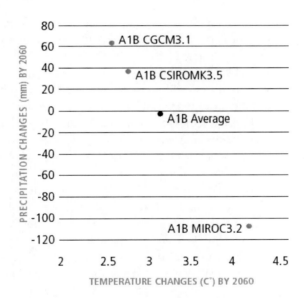

Figure 6. U.S. temperature and precipitation changes from 1961 to 1990 (historical period) to the decade surrounding the year 2060 (2055 to 2064) under a future of high economic growth and moderate population growth (A1B) and climate scenarios predicted by three general circulation models of climate change: CGCM3.1, CSIROMK3.5, and MIROC3.2. (source: 2010 Resources Planning Act assessment).

The Intergovernmental Panel on Climate Change climate projections first were downscaled to the approximately 10-km scale and then aggregated to the county scale (Joyce et al., in press). At the scale of the contiguous United States, the average annual temperatures and total annual precipitation under

A1B represent the warmest and the driest climate at 2060 (Fig. 6). Within A1B, the CGCM3.1 reflects the coolest (plus 2.55 °C) and wettest (plus 62.32 mm) climate over the projection period, and MIROC3.2 reflects the warmest (plus 4.21 °C) and driest (minus 107.39 mm); intermediate to them is CSIROMK3.5 with moderate temperature (plus 2.73 °C) and precipitation (plus 37.54 mm) changes. Although all areas of the United States show increases in temperature, the rates of change among regions vary somewhat, and regional differences in precipitation projections vary considerably (Joyce et al., in press).

Summary

The purpose of this report is to evaluate how changes in population, demographics, economic conditions, land use, and climate likely will affect participants and days of participation in the North for 17 natural resource-based recreation activities. The demographic, climate, and land use projections described above were used to develop projections of future resource uses and conditions. Not all of the projected variables were used in all models, but all of the projection models used some subset of these variables. As a result, the projections and their underlying assumptions provide a common framework for comparing results across three climate futures associated with A1B.

This report proceeds in three main parts. First, we present the statistical methods and previous research that underlie regional per capita participation and consumption models. Next, we describe the data used in the estimation step, including projections of the various income and population growth factors and relevant assumptions—and we present estimation and simulation steps for regional participation and days projections by activity and climate future for A1B to 2060. Finally, we discuss some of the key findings within and across categories, and in association with the factors that are likely to drive change over the projection period.

METHODS AND DATA

Models used to assess recreation demand decisions can be grouped into three basic categories: (1) site-specific user models, (2) site-specific aggregate models, and (3) population-level models. Cicchetti (1973) pioneered cross-sectional population-level models with the household-based 1965 National

Survey of Recreation, which estimated annual participation and use nationally for many outdoor recreation activities; then used estimated models and Census Bureau projections of sociodemographic variables and population to forecast participation and use to 2000. Researchers have used the cross-sectional population-level approach to estimate and project participation and use for recreation activities at national and regional levels (Bowker 2001; Bowker et al., in press; Hof and Kaiser 1983a; Leeworthy et al. 2005; Walsh et al. 1992) and for previous RPA assessments (Bowker et al. 1999, Bowker et al. 2012, Hof and Kaiser 1983b). Researchers also have used alternative approaches— combining population data with individual site-level data or county-level data—to project national or regional recreation demand (Bowker et al. 2006, Cordell and Bergstrom 1991, Cordell et al. 1990, Englin and Shonkwiler 1995, English et al. 1993, Poudyal et al. 2008).

A major drawback of cross-sectional models, imposed by the nature of the data, is that the structure of the estimated models remains constant over the forecast period.

For example, the factors that influence participation or use are assumed to have the same effects throughout the projection period. Hence, with model parameters constant in time and barring major shifts in demographics, the results often are driven by population growth. This assumption can be tenuous. One consequence is that new sports brought about by technological changes or shifts in tastes and preferences (such as mountain biking, jet skiing, snowboarding, flat-water kayaking, and orienteering) are less likely to be correctly represented in the models. Moreover, if data are collected while activities are in a new or rapid growth phase, recent trends can be misleading; for example, although Cordell (2012) reported a recent increase in kayaking participation of 154 percent in less than a decade, sustaining such a rate of growth for 50 years would be unlikely. A further drawback of these models is the difficulty of accounting for the dampening effects of future congestion, supply limitations, economic downturns or upswings, and relative price changes on growth in participation and use. Nevertheless, without appropriate time-series data or panel data, researchers are left with the inherent limitations of cross-sectional models, as a second-best alternative to estimate and forecast participation and use.

Regional cross-sectional population-level logistic models were used to describe the probability of adult participation in each of the 17 activities as:

$$P_i = \frac{1}{[1+\exp(-X_i\, B_i)]}$$

(eq. 1)

where

Pi = the probability that an individual participated in recreation activity i in the preceding year

Xi = a vector that contains sociodemographic characteristics unique to activity i across individuals, relevant supply variables for activity i across individual locations (Table 3), and at least one climate variable related to conditions at or near the individual's residence

Bi = a vector of parameters associated with activity i

Data were manipulated using SAS 9.1 (2004) and models were estimated using NLOGIT 4.0 (Greene 2009).

Logistic models for each activity, based on NSRE data from 1999 to 2008 (U.S. Department of Agriculture Forest Service 2009), were combined with 2008 baseline population-weighted sample averages for the explanatory variables to create an initial predicted per capita participation rate for each activity. The per capita participation rates were recalculated at 10-year intervals using projected changes in the explanatory variables. Indices then were created for the participation rates by which the NSRE 2005 to 2009 average population-weighted participation frequencies (Table 1) were scaled, leading to indexed per capita participation rates for each of the 17 activities.

We opted to index the NSRE averages by changes in model-predicted rates because doing so mitigates the potential for nonlinearity biases that are associated with complete reliance on logistic predicted values (Souter and Bowker 1996). The indexed participation rate estimates then were combined with projected changes in population, according to each of the three 2010 RPA assessment scenarios, to yield indexed values for total adult participants in the region across the 17 activities.

Participation intensity or consumption models were similar to the participation models listed above except that an integer metric represented use—the number of times, days, visits, trips, or events was modeled rather than the binary (yes/no) decision to participate. The general specification for the population-level consumption model was:

$$Y_i = f(X_i,\, Q_i) + u_i$$

(eq. 2)

where

Yi = the annual number of different days during which an individual participates in activity i

Xi = a vector of individual sociodemographic characteristics

Qi = a vector of supply relevant variables for activity i

ui = a random disturbance term specific to activity i

These integer or count data models are often estimated using negative binomial specifications with a link function of semi-logarithmic form (Bowker 2001, Bowker et al. 1999, Zawacki et al. 2000).

Variations of these consumption/demand models have been developed for onsite applications, where all observed visits were recorded as positive integers (Bowker and Leeworthy 1998). Such zero-truncated models have been applied extensively in onsite recreation demand estimation and valuation research (Bowker et al. 2007, Englin and Shonkwiler 1995, Hagerty and Moeltner 2005, Ovaskainen et al. 2001). In some studies, the estimated models have been extrapolated to general populations, assuming that visitors and nonvisitors come from the same general population of users (Englin and Shonkwiler 1995). This approach, wherein population data are combined with individual site-level data, was suggested by Cordell and Bergstrom (1991) and used in a previous RPA assessment by Cordell et al. (1990) with linear models to estimate outdoor recreation trips for 31 activities and to project the number of trips by activity from 1989 to 2040. English et al. (1993) extended the RPA models and projections to the regional level by combining parameter estimates from national models with regional explanatory variable values. Among others, Hagerty and Moeltner (2005) question the efficacy of extrapolating parameter estimates from the onsite demand models to the population at large.

Household data, such as from the NSRE, may report zero visits; doing so eliminates problems related to onsite samples and extrapolating onsite models to general populations. In a previous RPA assessment analysis, Bowker et al. (1999) used data from the 1994 to 1995 NSRE, the U.S. Census, and the 1997 Forest Service National Outdoor Recreation Supply Information System database to project of participation and consumption (annual days and trips) for more than 20 natural resource-based outdoor activities, both nationally and in four geographical regions of the United States, from 2000 to 2050. That analysis moved beyond participation modeling to include negative binomial count models to estimate consumption (days and trips annually) and to project these measures over the same time period.

Table 3. Socioeconomic and supply variables for modeling and forecasting outdoor recreation participation and days-of-participation by American adults

variable	description
Gender	1=Male, 0=otherwise
American Indian	1=American Indian, 0=otherwise
Asian/Pacific Islander	1=Asian/Pacific Islander, 0=otherwise
Hispanic	1=Hispanic, 0=otherwise
Black	1=African American, 0=otherwise
Bachelors	1=Bachelor degree, 0=otherwise
Below High school	1=Less than high school, 0=otherwise
Post Graduate	1=Post-graduate degree, 0=otherwise
some College	1=some college or technical school, 0=otherwise
Age	respondent age in years
Age squared	respondent age squared
Income	respondent household income (2007 dollars)
Population density	County area divided by population (base 1997)
Coastal	1=County on coast, 0=otherwise
For_ran_pcap	sum of forest land acres and rangeland acres divided by population at county level and at 50, 100, 200-mile radii (base 1997)
Water_pcap	Water acres divided by population at county level and at 50, 100, 200-mile radii (base 1997)
Mtns_pcap	Mountainous acres divided by population (base 1997)
Pct_mtns_pcap	Percentage of county acres in mountains divided by population multiplied by 100000 (base 1997)
natpark_pcap	number of nature parks and similar institutions divided by population multiplied by 100000 (base 1997)
Fed_land_pcap	sum u.s. Forest service, national Park service, u.s. Fish and Wildlife service, Bureau of Land Management, u.s. Bureau of reclamation, tennessee Valley Authority, and u.s. Army Corps of engineers acreage divided by population (base 1997)
Avg_elev	Average elevation in meters at county level and 50-, 100-, 200-mile radii (base 1997)

Bowker (2001) followed the same approach, using NSRE and State-level data in projections from 2000 to 2020 of outdoor-recreation participation and consumption in Alaska. Leeworthy et al. (2005) used NSRE 2000 data in projecting participation and consumption of marine-related outdoor recreation through 2010. Bowker et al. (2006) applied similar methods with 2000 NSRE and National Visitor Use Monitoring data (English et al. 2002, U.S. Department of Agriculture Forest Service 2010) in developing projections of wilderness and primitive area recreation participation and consumption from 2002 through 2050.

Alternatively, if one suspects that observed zeros for the dependent variable (days of participation) are excessive or not entirely caused by the same data generating process as the positive values, using a hurdle model structure or a zero-inflated count procedure would be appropriate (Cameron and Trivedi 1998). The hurdle model that we employed combines the probability of participation (threshold) with the estimated number of days for those participating, as shown below.

$$E[Y|X] = Pr[Y > 0|X1] * E_{y>0}[Y|Y > 0, X2]$$

(eq. 3)

where
    E = expected value operator
    Pr = probability of participation
    Y = days of participation
    X = vector of explanatory variables.

The hurdle model allowed different vectors of explanatory variables ($X1$ and $X2$), and thus parameters, for the respective products of the expectation in eq. 3—probability and conditional-days portions of the model—with probability estimated as a logistic (eq. 1) and conditional days estimated as a truncated negative binomial, thus leading to two unique sets of estimated parameters. Each of the 17 regional outdoor recreation activity day hurdle models were estimated with NLOGIT 4.0 (Greene 2009), using 1999 to 2009 NSRE data for U.S. households (U.S. Department of Agriculture Forest Service 2009), county-level climate data (Joyce et al., in press), county-level land use data (Wear 2011), and recreation supply data (Cordell et al. 2012). Although we did not formally test the hurdle model against the simpler untruncated negative binomial model (Bowker et al. 1999) for each activity, the parameter estimates and the significant variables for the logistic portion nearly always differed from the conditional days portion, thus validating the choice of the hurdle model.

As in the procedure for the participation models and indices, hurdle model parameter estimates were combined with 2008 NSRE baseline participation and days estimates (Table 1), projected explanatory variables, and projected population changes under each of the climate scenarios to provide indices of annual days-ofparticipation growth projections for the activities listed in Table 1. The three climate scenarios (CGCM3.1, MIROC3.2, and CSIROMK3.5) were used in conjunction with A1B.

Table 3 lists socioeconomic and supply variables for the various models and projections. The preponderance of these variables was included in the NSRE database (U.S. Department of Agriculture Forest Service 2009). Additional variables related to supply were obtained from Cordell et al. (2013). Projections of land use change variables were obtained from Wear (2011). Historical as well as projected climate data were obtained from Joyce et al. (in press). As little or no literature was available to link climate with household participation and consumption of recreation activities, an ad hoc approach was followed during the model estimation stage, wherein climate variables were created based on 6-year moving averages and arbitrary distances from county centroids. Table 4 lists representative climate variables. Each estimated model was limited to one climate variable; selection occurred on an ad hoc basis, primarily based on model fit.

Table 4. Climate variables used for estimating and forecasting outdoor recreation participation and days-ofparticipation by American adults

variable	description
Ppt_monthly_ average_d100[a]	daily average of precipitation for all months for resident county and counties within 100 miles of resident county centroid
Ppt_monthly_ average_d200	daily average of precipitation for all months for resident county and counties within 200 miles of resident county centroid
spring_Pet_d200	spring average daily potential evapotranspiration for resident county and counties within 200 miles of resident county centroid
tmax_fall_d50	Average monthly maximum fall temperature for resident county and counties within 50 miles of resident county centroid
tmax_geq_25_d200	Percentage of month where average monthly maximum temperature exceeded 25 degrees Celsius for resident county and counties within 200 miles of resident county centroid
tmax_geq_35	Percentage of months where average monthly maximum temperature exceeded 35 degrees Celsius in the resident county

variable	description
tmax_geq35_d100	Percentage of month where average monthly maximum temperature exceeded 35 degrees Celsius for resident county and counties within 100 miles of resident county centroid
tmax_geq35_d200	Percentage of month where average monthly maximum temperature exceeded 35 degrees Celsius for resident county and counties within 200 miles of resident county centroid
tmax_spring	Average of the average monthly maximum temperature in spring in the resident county
tmax_spring_d100	Average of the average monthly maximum temperature in spring for the resident county and counties within 100 miles of resident county centroid
tmax_summer	Average of the average monthly maximum temperature in summer in the resident county
tmax_summer_d50	Average of the average monthly maximum temperature in summer for the resident county and counties within 50 miles of resident county centroid
tmax_summer_d100	Average of the average monthly maximum temperature in summer for the resident county and counties within 100 miles of resident county centroid
tmax_summer_d200	Average of the average monthly maximum temperature in summer for the resident county and counties within 200 miles of resident county centroid
tmax_winter	Average of the average monthly maximum temperature in winter in the resident county
tmax_winter_d100	Average of the average monthly maximum temperature in winter for the resident county and counties within 100 miles of resident county centroid
tmin_leq_0	Percent of month where average monthly minimum temperature was below 0 degrees Celsius in the resident county
tmin_leq_neg10	Percent of month where average monthly minimum temperature was below -10 degrees Celsius in the resident county
total_ppt_d100	Monthly average of total monthly precipitation in resident county and counties within 100 miles of resident county centroid
total_ppt_d200	Monthly average of total monthly precipitation in resident county and counties within 200 miles of resident county centroid
Winter_Pet_d50	Average of average daily potential evapotranspiration in winter for resident county and counties within 50 miles of resident county centroid

Table 4. (Continued)

variable	description
Winter_Pet_d200	Average of average daily potential evapotranspiration in winter for resident county and counties within 200 miles of resident county centroid
Yearly_Pet_d200	Average of average daily potential evapotranspiration for resident county and counties within 200 miles of resident county centroid

[a] All averages are calculated over 6-year periods, for example, historic data are based on 2001 to 2006 data, 2060 projections are based on averages from 2055 to 2060. Seasons were divided into 3-month periods based on the following categories: winter (December, January, and February), spring (March, April, and May), summer (June, July, and August), and fall (September, October, and November).

RESULTS

As discussed in the previous section, results were estimated for 17 outdoor recreation participation models (eq. 1). All models included socioeconomic variables, in addition to at least one variable reflecting land use change or relative supply of settings typically associated with the particular activity. In addition, each model included one climate variable (Table 5). Reported results include parameter estimates for each activity, values for explanatory variables by scenario and year, odds ratios which indicate the probability of participation occurring in one group compared to the probability of it occurring in another group, fit statistics, and graphics of total participant growth by activity and climate scenario.

Logistic parameter estimates then were combined with available projections of relevant explanatory variables under the updated A1B storyline (high economic growth, moderate population growth), including one associated with each of the three climate scenarios—the coolest and wettest CGCM3.1, the warmest and driest MIROC3.2, and the intermediate CSIROMK3.5—to create indexed per capita participation estimates at 10-year intervals through 2060. These indices were, in turn, combined with population projections for A1B to develop estimated participant indices.

The participant indices then were applied to a beginning baseline estimate of participants for each activity (based on weighted regional averages calculated from 2005 to 2009 NSRE data) to yield the projected number of adult participants. The 4-year average around 2008 was chosen to avoid any abnormality associated with a single year.

The hurdle model (eq. 3) combined the probability of participation in an activity with the expected value of days participating, given that the individual actually participated. The estimated logistic models were thus combined with conditional participation days models to complete the hurdle model. Given that only those who participated were included in the conditional days portion of the model (thus ensuring no zero observations for days), a truncated negative binomial model was employed for estimation. As with the logistic participation models above, days models were estimated for each of the 17 outdoor recreation activities reported in Table 1. Table 5 lists climate variables used in the days models.

Table 5. Climate variables used for estimating and forecasting outdoor recreation participation and days-of-participation

recreation activity	Model	climate variable
developed site use	Participation days	summer_Pet_d200 spring_Pet
Interpretive site use	Participation days	total_ppt_d100 Fall_Pet
Birding	Participation days	spring_Pet_d200 total_ppt_d50
nature viewing	Participation days	summer_Pet_d200 total_ppt_d50
Challenge	Participation days	total_ppt total_ppt_d50
equestrian	Participation days	summer_Pet_d200 tmin_fall
day hiking	Participation days	Winter_Pet_d200 total_ppt_d50
Primitive area use	Participation days	Fall_Pet_d200 summer_Pet_d50
off-road driving	Participation days	tmax_fall_d200 total_ppt_d200
Motorized water	Participation days	summer_Pet_d50 total_ppt_d200
Motorized snow	Participation days	Winter_Pet_d200 total_ppt_d100
Hunting	Participation days	Fall_Pet_d100 tmax_geq_25_d100
Fishing	Participation days	summer_Pet_d100 summer_Pet_d200
developed skiing	Participation days	total_ppt spring_Pet_d200
undeveloped skiing	Participation days	tmax_winter_d200 tmax_winter_d200
swimming	Participation days	tmax_summer_d200 tmax_geq_25_d200
Floating	Participation days	summer_Pet_d200 spring_Pet_d200

Total days for each activity were estimated following a procedure similar to that for estimating participants. First, days of participation per participant were nonlinearly regressed on relevant explanatory variables including at least one climate variable. Parameter estimates from the respective negative binomial models then were combined with projected explanatory variables under A1B and associated climate model forecasts, at 10-year intervals, to create indexed per capita days of participation, which were combined with population projections for A1B to develop estimated per participant days indices. These indices then were applied to a beginning baseline estimate of participation days for each activity, based on weighted national averages calculated from the 2005 to 2009 NSRE data, to yield projections of total adult activity days. As with the participant estimates, a 4-year average around 2008 was chosen to avoid any aberrations associated with a single year. The results of participation and days-of-participation models are shown in the series of tables that follow. In addition to results simulated under each climate scenario associated with A1B, an average across the climate scenarios is also reported.

Developed Site Use

Visiting Developed Sites
Activities associated with this composite activity include family gathering, picnicking, and developed camping. Per capita participation for this activity is currently high and projected to remain relatively constant into the future, decreasing slightly on average, across climate scenarios (Table 6). The minor decrease in adult participation rate, coupled with population growth, suggests an increase in users of 27 to 35 percent by 2060, or about 24 million more per year than the current 80.5 million.

Average annual days per developed site visitor are projected to decrease by 10 percent on average across climate scenarios, or just over 1 day per participant per year. CSIRO and MIROC climate forecasts roughly doubled the decrease from the CGCM model (Table 6). Total days of developed use increase 12 to 27 percent across the climate scenarios. Given the relatively small changes in average days of developed site visitation per participant across the climate scenarios, the key driver in the increase in total days for this activity is likely to be population growth. Across all climate scenarios, the average expected increase in annual days of developed site visitation is about 17 percent or 160 million days for the region annually by 2060.

Table 6. Projected developed site visit participation and use (family gathering, picnicking, developed camping) by adults in the Northern United States, 2008 to 2060, under Resources Planning Act (RPA) scenario A1B and related climate futures

RPA scenario	year					
	2008	2060 climate average[a]	2060 climate average[a]	2060 CGCM3.1	2060 CSIRO-MK3.5	2060 MIROC3.2
	Per capita participation	Percent increase (decrease) from 2008				
A1B	0.825	0.791	(4)[b]	0	(6)	(6)
	Adult participants (millions)	Percent increase (decrease) from 2008				
A1B	80.5	104.2	30	35	27	27
	days per participant	Percent increase (decrease) from 2008				
A1B	11.69	10.58	(10)	(5)	(11)	(12)
	total days (millions)	Percent increase (decrease) from 2008				
A1B	943.3	1,106.2	17	27	13	12

[a] Result of average across CGCM3.1, CSIROMK3.5, and MIROC3.2.
[b] Parentheses denotes decrease.

Visiting Interpretive Areas

Interpretive areas include nature centers, zoos, historic sites, and prehistoric sites. More than 67 million adults, or about 69 percent of all residents in the region, participated in at least one activity in this category annually from 2005 to 2009. The projections indicate participation rates are likely to be stable at about 9 percent across climate scenarios (Table 7). Climate effects are expected to result in small differences in participation rates, offset by consistently higher numbers of days per participant (up to half a day per year on average). As per capita participation is expected to rise 9 percent, the number of participants will likely exceed the rate of population growth. The higher growth in participation rate for this activity group compared to visiting developed sites has several possible causes—visiting developed sites is negatively related to age, which is expected to rise by 2060, and positively related to available Federal land per capita—that are less important in interpretive site visitation, as is climate change. Total annual days of interpretive site visits is projected to increase by two-thirds, on average, or by 343 million days per year by 2060.

Table 7. Projected interpretive site visit participation and use (visiting natural sites, prehistoric, or historic sites) by adults in the Northern United States, 2008 to 2060, under Resources Planning Act (RPA) scenario A1B and related climate futures

RPA scenario	year					
	2008	2060 climate average[a]	2060 climate average[a]	2060 CGCM3.1	2060 csiro-MK3.5	2060 MIROC 3.2
	Per capita participation	Percent increase (decrease) from 2008				
A1B	0.686	0.747	9	9	9	9
	Adult participants (millions)	Percent increase (decrease) from 2008				
A1B	67.0	98.5	47	47	48	47
	days per participant	Percent increase (decrease) from 2008				
A1B	7.69	8.70	13	8	14	17
	total days (millions)	Percent increase (decrease) from 2008				
A1B	516.0	858.5	66	59	68	72

[a] Result of average across CGCM3.1, CSIROMK3.5, and MIROC3.2.

Observing Nature

This category includes birding, both viewing and photographing; it also includes the more general activity aggregate called viewing, which consists of any activities that involve the viewing, photography, or study of natural settings, or the noncommercial gathering of plants or animals. From 2005 to 2009, an average of 38 percent of northern adults, or 37 million people, participated annually in birding. In the more broadly defined viewing aggregate, which would include birding, nearly 82 percent of the adult population, or about 79 million people, participated annually.

Birding
Participation in birding is expected to remain stable over the next 50 years, with the participation rate declining by about 1.4 percent to about 37

percent of northern adults. Coupling this decrease with the population growth expected under A1B would mean a regional increase in birders of 24 to 40 percent, depending on the climate scenario (Table 8). On average, the expected annual increase is 30 percent, or about 11 million adults in 2060. The number of days per participant is expected to decrease uniformly (4 percent) across the three climate scenarios. Given that adult birders in the region averaged nearly 100 days per year from 2005 to 2009, an annual decrease of 4 days would not have much of an effect on the annual totals, which should increase by 19 to 35 percent over the 50-year period. The largest increase, 1,310 million days per year, would occur under CGM3.1, which is marginally wetter and cooler than the other two scenarios at the national level.

Table 8. Projected birding participation and use (viewing or photographing birds) by adults in the Northern United States, 2008 to 2060, under Resources Planning Act (RPA) scenario A1B and related climate futures

	year					
RPA scenario	2008	2060 climate average[a]	2060 climate average[a]	2060 CGCM3.1	2060 CSIRO-MK3.5	2060 MIROC 3.2
	Per capita participation		Percent increase (decrease) from 2008			
A1B	0.382	0.368	(4)b	3	(7)	(8)
	Adult participants (millions)		Percent increase (decrease) from 2008			
A1B	37.2	48.3	30	40	26	24
	days per participant		Percent increase (decrease) from 2008			
A1B	99.8	96.1	(4)	(3)	(4)	(4)
	total days (millions)		Percent increase (decrease) from 2008			
A1B	3,696	4,625	25	35	21	19

[a] Result of average across CGCM3.1, CSIROMK3.5, and MIROC3.2.
[b] Parentheses denotes decrease.

Viewing

The regional adult participation rate in the broader viewing category will likely remain essentially unchanged over the next 50 years, suggesting that viewing participants will increase at about the rate of population increase. By 2060, the total number of nature viewers per year is expected to increase by 35 percent to about 107 million adults (Table 9). Annual average nature viewing days per participant will likely decrease across all scenarios by 8 to 10 percent, or about 2 weeks per year, resulting in one of the largest relative decreases in total days per participant across all activities (Table 9). The decrease in viewing days per participant appears to be driven by a number of factors, among them, projected increases in total population density and in minority populations, and a projected decrease in public land per capita in the region. Despite the predicted decrease in annual days per participant, total viewing days will likely increase, driven by the increase in the number of participants, by an average of about 3,104 million days per year by 2060.

Table 9. Projected nature viewing participation and use (viewing, photography, study, or nature gathering related to fauna, flora, or natural settings) by adults in the Northern United States, 2008 to 2060, under Resources Planning Act (RPA) scenario A1B and related climate futures

	year					
RPA scenario	2008	2060 climate average[a]	2060 climate average[a]	2060 CGCM3.1	2060 CSIRO-MK3.5	2060 MIROC 3.2
	Per capita participation	Percent increase (decrease) from 2008				
A1B	0.815	0.813	0	3	(2)b	(2)
	Adult participants (millions)	Percent increase (decrease) from 2008				
A1B	79.5	107.0	35	39	32	33
	days per participant	Percent increase (decrease) from 2008				
A1B	175.7	159.5	(9)	(8)	(10)	(9)
	total days (millions)	Percent increase (decrease) from 2008				
A1B	13,925	17,029	22	27	19	21

[a] Result of average across CGCM3.1, CSIROMK3.5, and MIROC3.2.
[b] Parentheses denotes decrease.

Backcountry Activities

Backcountry activities are most often pursued in undeveloped but accessible lands. The category includes these four activities, or activity composites: (1) challenge activities, (2) horseback riding, (3) hiking, and (4) visiting primitive areas.

Challenge Activities

Challenge activities, often associated with young and affluent adults, include caving, mountain biking, mountain climbing, and rock climbing.

Outlook for Outdoor Recreation in the Northern United States   95

Nearly 10 percent of adults in the region currently engage in these activities, a rate expected to decrease by about 10 percent in 50 years (Table 10).

Table 10. Projected challenge activity participation and use (mountain climbing, rock climbing, caving) by adults in the Northern United States, 2008 to 2060, under Resources Planning Act (RPA) scenario A1B and related climate futures

RPA scenario	year					
	2008	2060 climate average[a]	2060 climate average[a]	2060 CGCM3.1	2060 CSIRO-MK3.5	2060 MIROC 3.2
	Per capita participation	Percent increase (decrease) from 2008				
A1B	0.095	0.086	(10)[b]	(6)	(14)	(9)
	Adult participants (millions)	Percent increase (decrease) from 2008				
A1B	9.4	11.4	22	27	16	22
	days per participant	Percent increase (decrease) from 2008				
A1B	3.89	3.82	(2)	0	(4)	(2)
	total days (millions)	Percent increase (decrease) from 2008				
A1B	37.7	45.1	20	27	12	21

[a] Result of average across CGCM3.1, CSIROMK3.5, and MIROC3.2.
[b] Parentheses denotes decrease.

Population growth in the region will likely offset expected participation-rate decreases, leading to increases in the number of participants of 16 to 27 percent across the climate scenarios. Participation is projected to grow by 22 percent on average, or by about 2 million adults per year through 2060. The number of days per participant will be almost unchanged across climate scenarios, remaining at less than 4 days per year among participants. Coupled with population growth rates, total days of challenge sport participation will likely increase 12 to 27 percent annually by 2060. On average this increase would result in an additional 7.4 million days of activity per year.

Horseback Riding

Horseback riding on trails claimed 6 percent of the northern adult population annually as participants in 2008— a percentage expected to increase to more than 8 percent by 2060, with the biggest increases occurring under the warmer and drier climate scenarios predicted by CSIROMK3.5 and MIROC3.2 (Table 11). When population growth is included to derive the number of annual participants, the expected average increase across the three climate scenarios is 91 percent, or just about twice the 5.8 million 2008 participants. Relative to previously discussed activities—such as visiting developed sites, nature viewing, and challenge sports—expected changes in climate do not appear to have a dampening effect on horseback-riding participation. The per capita number of days of participation would remain about constant over the projection period, dropping about a half day per year, or 4 percent on average. However, factoring in population growth would lead to increases in the total days of horseback riding of 65 to 100 percent by 2060, depending on the climate scenario, with the average increase expected to be about 61 million days per year.

Table 11. Projected equestrian participation and use (horseback riding on trails) by adults in the Northern United States, 2008 to 2060, under Resources Planning Act (RPA) scenario A1B and related climate futures

RPA scenario	year					
	2008	2060 climate average[a]	2060 climate average[a]	2060 CGCM3.1	2060 CSIRO-MK3.5	2060 MIROC 3.2
	Per capita participation		Percent increase (decrease) from 2008			
A1B	0.059	0.084	42	23	54	48
	Adult participants (millions)		Percent increase (decrease) from 2008			
A1B	5.77	11.05	91	67	108	100
	days per participant		Percent increase (decrease) from 2008			
A1B	12.63	12.16	(4)[b]	(1)	(4)	(7)
	total days (millions)		Percent increase (decrease) from 2008			
A1B	72.3	133.0	84	65	100	87

[a] Result of average across CGCM3.1, CSIROMK3.5, and MIROC3.2.
[b] Parentheses denotes decrease.

Day Hiking

Hiking is the most popular single backcountry activity with about a third of all northern adults, or about 32.4 million people, hiking in 2008. Among the three climate scenarios, hiking participation per capita is expected to remain about constant out to 2060 (Table 12). Thus, with population growth, hikers in the region should increase by about a third in 2060. Hiking is the only activity in which Hispanics demonstrated a higher participation rate than Caucasians. Annual days of hiking per participant are forecasted to decrease evenly across the climate scenarios, averaging 7 percent or about 1.6 days per year.

Thus, total annual days of hiking will likely increase less than population growth, but the result would nevertheless be an increase in hiking days of approximately 162 million days by 2060.

Table 12. Projected day hiking participation and use by adults in the Northern United States, 2008 to 2060, under Resources Planning Act (RPA) scenario A1B and related climate futures

RPA scenario	year					
	2008	2060 climate average[a]	2060 climate average[a]	2060 CGCM3.1	2060 CSIRO-MK3.5	2060 MIROC 3.2
	Per capita participation		Percent increase (decrease) from 2008			
A1B	0.327	0.319	(2)b	0	(5)	(2)
	Adult participants (millions)		Percent increase (decrease) from 2008			
A1B	32.4	42.7	32	35	28	32
	days per participant		Percent increase (decrease) from 2008			
A1B	22.44	20.86	(7)	(6)	(9)	(7)
	total days (millions)		Percent increase (decrease) from 2008			
A1B	723.8	886.0	22	28	17	23

[a] Result of average across CGCM3.1, CSIROMK3.5, and MIROC3.2.
[b] Parentheses denotes decrease.

Visiting Primitive Areas

The final backcountry activity, an aggregate called visiting primitive areas, consists of backpacking, primitive camping, and visiting a designated or undesignated wilderness. This composite accounted for 36.1 million regional participants in 2008, or about 37 percent of all adults. Annual per capita participation in this category is expected to decrease 10 to 34 percent over the next 50 years across the climate scenarios, an average 8.5-percent drop (Table 13). Increased population density, declining Federal land area per capita, and increasing population diversity appear to be factors influencing the participation rate decrease. However, overall participation is expected to increase by an average of 4 percent, to under 38 million adults by 2060, because population growth offsets the decrease in participation rates.

Average annual days per participant visiting primitive areas is projected to decrease 14 to 26 percent across climate scenarios (Table 13), to more than 2 days per year by 2060. The decrease in participation rate and the drop in average participant days per year would lead to annual average decreases of 18 percent, from the current 415 million to 342 million days per year. However, the climate scenario showing the smallest change from the 2008 baseline, CGCM3.1, predicts a 5-percent annual increase in visitation days to less than or equal to 436 million by 2060.

Table 13. Projected primitive area visit participation and use (backpacking, primitive camping, wilderness) by adults in the Northern United States, 2008 to 2060, under Resources Planning Act (RPA) scenario A1B and related climate futures

RPA scenario	year					
	2008	2060 climate average[a]	2060 climate average[a]	2060 CGCM3.1	2060 CSIRO-MK3.5	2060 MIROC 3.2
	Per capita participation		Percent increase (decrease) from 2008			
A1B	0.367	0.282	(23)b	(10)	(26)	(34)
	Adult participants (millions)		Percent increase (decrease) from 2008			
A1B	36.1	37.4	4	22	0	(11)
	days per participant		Percent increase (decrease) from 2008			
A1B	11.42	9.01	(21)	(14)	(26)	(24)
	total days (millions)		Percent increase (decrease) from 2008			
A1B	415	342.0	(18)	5	(26)	(32)

[a] Result of average across CGCM3.1, CSIROMK3.5, and MIROC3.2.
[b] Parentheses denotes decrease.

Motorized Activities

Three categories of motorized activities were considered: off-road driving, motorized water use, and snow use.

Off-Road Driving

Participation in off-road driving averaged about 18 percent of the northern adult population, or about 17.3 million adults, annually from 2005 to 2009 (Table 14). Future participation rates are expected to decrease by 4 to 13 percent, depending on the climate scenarios. Among factors leading to the decrease are the expected increase in minority populations and general aging of the total population. Despite these declining rates of growth in per capita participation, the number of participants in off-road driving will likely increase 18 to 29 percent under the climate scenarios to somewhere between 20 and 22 million people in 2060, because the rate of population growth is expected to

Outlook for Outdoor Recreation in the Northern United States   99

outstrip any decrease in per capita participation. Annual days of off-road driving per participant is projected to decrease 11 to 14 percent, or about 2 days per year by 2060 (Table 14) with only small variations among climate scenarios.

Table 14. Projected motorized off-road participation and use (off-road driving) by adults in the Northern United States, 2008 to 2060, under Resources Planning Act (RPA) scenario A1B and related climate futures

RPA scenario	year					
	2008	2060 climate average[a]	2060 climate average[a]	2060 CGCM3.1	2060 CSIRO-MK3.5	2060 MIROC 3.2
	Per capita participation		Percent increase (decrease) from 2008			
A1B	0.176	0.162	(8)[b]	(4)	(8)	(13)
	Adult participants (millions)		Percent increase (decrease) from 2008			
A1B	17.3	21.4	24	29	25	18
	days per participant		Percent increase (decrease) from 2008			
A1B	16.43	14.36	(13)	(11)	(14)	(13)
	total days (millions)		Percent increase (decrease) from 2008			
A1B	282.8	306.7	8	16	7	3

[a] Result of average across CGCM3.1, CSIROMK3.5, and MIROC3.2.
[b] Parentheses denotes decrease.

These decreases in participation rate and average annual days per participant imply that, under all scenarios, the total number of days of off-road driving will increase at a lower rate than respective population growth rates. Nevertheless, on average, the amount of total off-road driving days per year is expected to increase from 282.8 to more than 306 million days in the region.

Motorized Snow Use

Motorized snow use, or snowmobiling, is a geographically limited activity undertaken by more than 7 percent of northern residents in 2008. Per capita participation in snowmobiling is projected to decrease 58 to 78 percent under all climate scenarios (Table 15). Regional changes in ethnicity, an aging population, declining Federal land per capita, and climate appear to be driving factors. Total snowmobiling participants are projected to decrease from 7 million in 2008 to between 2.1 and 3.9 million by 2060, depending on the climate scenario. Average annual days per participant would decrease by about 1 day per year on average. Coupled with the decrease in numbers of participants, this suggests a potential decrease in annual snowmobiling days of

52 to 74 percent by 2060. Averaged across the climate scenarios, the change implies a drop of annual snowmobiling days from 54.8 million in 2008 to 20.3 million in 2060.

Table 15. Projected motorized snow activity participation and use (snowmobiling) by adults in the Northern United States, 2008 to 2060, under Resources Planning Act (RPA) scenario A1B and related climate futures

RPA scenario	year					
	2008	2060 climate average[a]	2060 climate average[a]	2060 CGCM3.1	2060 CSIRO-MK3.5	2060 MIROC 3.2
	Per capita participation		Percent increase (decrease) from 2008			
A1B	0.071	0.022	(69)[b]	(58)	(78)	(69)
	Adult participants (millions)		Percent increase (decrease) from 2008			
A1B	7.0	3.0	(58)	(44)	(70)	(59)
	days per participant		Percent increase (decrease) from 2008			
A1B	7.87	6.89	(12)	(14)	(11)	(12)
	total days (millions)		Percent increase (decrease) from 2008			
A1B	54.8	20.3	(63)	(52)	(74)	(64)

[a] Result of average across CGCM3.1, CSIROMK3.5, and MIROC3.2.
[b] Parentheses denotes decrease.

Table 16. Projected motorized water participation and use (motor boating, waterskiing, using personal watercraft) by adults in the Northern United States, 2008 to 2060, under Resources Planning Act (RPA) scenario A1B and related climate futures

RPA scenario	year					
	2008	2060 climate average[a]	2060 climate average[a]	2060 CGCM3.1	2060 CSIRO-MK3.5	2060 MIROC 3.2
	Per capita participation		Percent increase (decrease) from 2008			
A1B	0.268	0.361	35	20	45	39
	Adult participants (millions)		Percent increase (decrease) from 2008			
A1B	26.1	47.3	82	61	96	88
	days per participant		Percent increase (decrease) from 2008			
A1B	14.65	16.01	9	6	12	10
	total days (millions)		Percent increase (decrease) from 2008			
A1B	378.8	752.8	99	72	118	106

[a] Result of average across CGCM3.1, CSIROMK3.5, and MIROC3.2.

Outlook for Outdoor Recreation in the Northern United States   101

Motorized Water Use

Motorized water activities involve motor boats, water skis, or personal watercraft. This combination of related activities had the highest per capita participation rate among motorized outdoor activities at 27 percent, or about 26.1 million adult participants, in 2008 (Table 16). Per capita participation is expected to grow by 20 to 45 percent over the next five decades to an average of 36 percent of all adults in the region. The highest growth rate is expected under the climate scenarios that are characterized by relatively higher average temperatures and less average rainfall. Overall, the numbers of adult participants in motorized water activities will likely increase faster than the population under all climate scenarios, for a total of 42 to 51 million participants in 2060.

Motorized water use participant days totaled about 378.8 million in 2008, or slightly less than 15 days annually per participant (Table 16). Days per participant are expected to increase 9 percent on average, or about 1.4 days per year by 2060. Combining population growth with increasing participation and annual days per participant would result in a 72- to 118-percent increase by 2060, meaning that, on average, the number of motorized water use days would double by 2060.

Hunting and Fishing

The traditional consumptive wildlife pursuits of hunting and fishing remain popular outdoor activities for northern adults, with about 11.3 million hunting and 28.7 million fishing participants in 2008. However, on a per capita basis, these pursuits have shown some decrease from past decades (Cordell 2012).

Hunting

Hunting is the legal pursuit of big game, small game, or migratory birds (as identified by an NSRE hunting screener question). The northern adult hunting participation rate, nearly 12 percent in 2008, is projected to decrease by 26 to 47 percent across climate scenarios by 2060 (Table 17)—with the pattern being that the bigger the change in climate conditions, the bigger the effect on hunting participation. The average decrease is projected to be 4 to 5 percentage points, meaning that about 7 percent of the adults will be hunting in 2060. The factors that appear to be associated with the drop in hunting participation are increased education levels, increased population density, diminishing availability of public land per capita, and increased minority populations.

Partly offset by population growth, the projected number of hunting participants is expected to drop on average across climate alternatives by 16 percent to about 9.5 million hunters by 2060. Across all the climate scenarios, average annual days in the field per hunter is projected to decrease 13 to 20 percent, or a little more than 3 days per hunter per year (Table 17). Climate appears to have less effect on the average annual days a hunter spends in the field than on whether one participates in hunting. Total annual adult hunting days, estimated at about 209.6 million in 2008, is expected to decrease by an average of about 30 percent by 2060 to just below 146 million days per year.

Table 17. Projected hunting participation and use (small game, big game, migratory bird, other) by adults in the Northern United States, 2008 to 2060, under Resources Planning Act (RPA) scenario A1B and related climate futures

	year					
RPA scenario	2008	2060 climate average[a]	2060 climate average[a]	2060 CGCM 3.1	2060 CSIRO-MK3.5	2060 MIROC 3.2
	Per capita participation		Percent increase (decrease) from 2008			
A1B	0.117	0.073	(38)[b]	(26)	(41)	(47)
	Adult participants (millions)		Percent increase (decrease) from 2008			
A1B	11.3	9.5	(16)	0	(20)	(28)
	days per participant		Percent increase (decrease) from 2008			
A1B	18.84	15.53	(18)	(13)	(20)	(19)
	total days (millions)		Percent increase (decrease) from 2008			
A1B	209.6	145.9	(30)	(12)	(36)	(42)

[a] Result of average across CGCM3.1, CSIROMK3.5, and MIROC3.2.
[b] Parentheses denotes decrease.

Fishing

Fishing participation—which includes warmwater and coldwater fishing, saltwater fishing, and anadromous fishing—can be either consumptive or catch-and-release. Unlike hunting, the adult participation rate for fishing is expected to increase over the next five decades. Currently, 29.6 percent of northern adults claim to fish. This rate is expected to increase by 3 to 27 percent by 2060 (Table 18). On average, the warmer and drier climate change scenarios, CSIROMK3.5 and MIROC3.2, would result in larger increases in

Outlook for Outdoor Recreation in the Northern United States 103

the participation rate than CGCM3.1. Coupled with population growth, the number of fishing participants is projected to rise by 39 to 71 percent by 2060. Averaged across climate scenarios, this implies an increase in annual anglers to about 45 million at the end of the projection period.

Fishing days per participant are forecasted to increase up to 19 percent by 2060, or by about 2 days per year on average (Table 18). Overall, annual fishing days are expected to increase across all climate scenarios by 39 to 104 percent during the next five decades, with the warmer and drier scenarios seeing the largest increases. On average, this would mean an increase in annual fishing days for the region of 78 percent, or about 403.6 million days.

Table 18. Projected fishing participation and use (cold water, warm water, saltwater, anadromous) by adults in the Northern United States, 2008 to 2060, under Resources Planning Act (RPA) scenario A1B and related climate futures

RPA scenario	year 2008	2060 climate average[a]	2060 climate average[a]	2060 CGCM 3.1	2060 CSIRO-MK3.5	2060 MIROC 3.2
	Per capita participation		Percent increase (decrease) from 2008			
A1B	0.296	0.347	17	3	27	22
	Adult participants (millions)		Percent increase (decrease) from 2008			
A1B	28.7	45.4	58	39	71	65
	days per participant		Percent increase (decrease) from 2008			
A1B	18.14	20.28	12	0	19	16
	total days (millions)		Percent increase (decrease) from 2008			
A1B	515.7	919.3	78	39	104	91

[a] Result of average across CGCM3.1, CSIROMK3.5, and MIROC3.2.

Non-Motorized Winter Activities

Non-motorized winter activities include developed skiing (downhill skiing and snowboarding) and undeveloped skiing (cross-country skiing and snowshoeing).

Developed Skiing

Developed skiing claimed an adult participation rate of 11.6 percent, about 11.6 million participants, annually from 2005 through 2009. Across the three climate scenarios, the participation rate for developed skiing is expected to increase by 25 to 32 percent or about 29 percent on average to about 15 percent of the adult population (Table 19). As with other income-dependent activities, the growth in household income associated with A1B would be a major driving factor in developed skiing participation rates, along with total precipitation, and education level increases. The increases in participation rate, combined with population growth, suggest that the number of developed skiing participants could grow by about 8.5 million participants to over 20 million per year by 2060.

Table 19. Projected developed skiing participation and use (downhill skiing, snowboarding) by adults in the Northern United States, 2008 to 2060, under Resources Planning Act (RPA) scenario A1B and related climate futures

	year					
RPA scenario	2008	2060 climate average[a]	2060 climate average[a]	2060 CGCM 3.1	2060 CSIRO-MK3.5	2060 MIROC 3.2
	Per capita participation		Percent increase (decrease) from 2008			
A1B	0.116	0.149	29	32	25	29
	Adult participants (millions)		Percent increase (decrease) from 2008			
A1B	11.6	20.1	74	78	69	75
	days per participant		Percent increase (decrease) from 2008			
A1B	6.99	4.75	(32)[b]	(22)	(37)	(37)
	total days (millions)		Percent increase (decrease) from 2008			
A1B	81.3	96.3	18	39	6	9

[a] Result of average across CGCM3.1, CSIROMK3.5, and MIROC3.2.
[b] Parentheses denotes decrease.

Alternatively, days of developed skiing per participant are projected to decrease by an average of 32 percent, or by about 2 days per participant annually, by 2060. This decrease is somewhat offset by population growth and the increase in number of participants, resulting in increased total skiing days annually across all three climate scenarios (Table 19). For climate scenario CGCM3.1, in which average annual temperature rises the least and average annual precipitation increases the most, the increase in total skiing days is 39 percent. For the equally likely warmer dryer scenarios, CSIROMK3.5 and MIROC3.2, the increases in total annual days are less than 10 percent, despite the large expected increases in participants.

Undeveloped Skiing

Undeveloped skiing often is pursued locally and does not require extensive recreation-site facilities. About 4.8 percent of northern adults, or 4.8 million people, engaged in undeveloped skiing in 2008. By 2060, this participation rate is projected to drop 27 to 41 percent, depending on the climate scenario (Table 20) with the warmer and drier MIROC3.2 showing the largest decrease. Other contributing factors include changing demographics in the region and declining public land per capita. Population growth would slightly offset the large decrease in participation rates, although on average the total number of participants in the region is projected to decrease by 11 percent, or about 0.6 million annually, by 2060.

Table 20. Projected undeveloped skiing (cross-country skiing, snowshoeing) by adults in the Northern United States, 2008 to 2060, under Resources Planning Act (RPA) scenario A1B and related climate futures

	year					
RPA scenario	2008	2060 climate average[a]	2060 climate average[a]	2060 CGCM 3.1	2060 CSIRO-MK3.5	2060 MIROC 3.2
	Per capita participation		Percent increase (decrease) from 2008			
A1B	0.048	0.032	(34)[b]	(35)	(27)	(41)
	Adult participants (millions)		Percent increase (decrease) from 2008			
A1B	4.8	4.2	(11)	(12)	(2)	(21)
	days per participant		Percent increase (decrease) from 2008			
A1B	6.66	6.09	(9)	(9)	(9)	(8)
	total days (millions)		Percent increase (decrease) from 2008			
A1B	32.1	26.0	(19)	(19)	(10)	(27)

[a] Result of average across CGCM3.1, CSIROMK3.5, and MIROC3.2.
[b] Parentheses denotes decrease.

Annual days per skier drop less than participation over the time period, less than 1 day per year, with little variation among climate scenarios. Thus, the predicted average 19-percent drop in undeveloped skiing days annually by 2060 appears to be primarily an artifact of the decreasing number of participants (Table 20). Overall, annual days of undeveloped skiing in the region are expected to decrease from 32.1 million to about 26 million by 2060.

Non-Motorized Water Activities

Non-motorized water activities consist of swimming and various forms of non-motorized boating.

Swimming

Swimming includes various related activities such as snorkeling, surfing, diving, and visiting beaches or watersides. It is the fourth most popular outdoor activity in the North, with a 63.3-percent adult participation rate and about 62 million adult participants annually (Table 21).

Table 21. Projected swimming participation and use (family gathering, picnicking, developed camping) by adults in the Northern United States, 2008 to 2060, under Resources Planning Act (RPA) scenario A1B and related climate futures

	year					
RPA scenario	2008	2060 climate average[a]	2060 climate average[a]	2060 CGCM 3.1	2060 CSIRO-MK3.5	2060 MIROC 3.2
	Per capita participation		Percent increase (decrease) from 2008			
A1B	0.633	0.633	0	9	(4)[b]	(4)
	Adult participants (millions)		Percent increase (decrease) from 2008			
A1B	61.7	83.3	35	47	29	29
	days per participant		Percent increase (decrease) from 2008			
A1B	22.24	21.13	(5)	2	(9)	(8)
	total days (millions)		Percent increase (decrease) from 2008			
A1B	1,376.2	1,771.8	29	51	17	18

[a] Result of average across CGCM3.1, CSIROMK3.5, and MIROC3.2.
[b] Parentheses denotes decrease.

On average, this participation rate is expected to remain constant over the projection period. Thus, the number of swimmers can be expected to increase by the same rate as the regional population. This would mean an increase of 17 to 29 million participants by 2060. Days per participant are projected to decrease slightly, 5 percent on average, under A1B and the three climate scenarios. Nevertheless, because of the high societal participation rate and the large number of days of annual engagement, swimming-related activities will likely increase 17 to 51 percent or by between 238 and 696 million days per year by 2060.

Table 22. Projected floating participation and use (canoeing, tubing, kayaking, rafting, sailing) by adults in the Northern United States, 2008 to 2060, under Resources Planning Act (RPA) scenario A1B and related climate futures

	year					
RPA scenario	2008	2060 climate average[a]	2060 climate average[a]	2060 CGCM 3.1	2060 CSIRO-MK3.5	2060 MIROC 3.2
	Per capita participation		Percent increase (decrease) from 2008			
A1B	0.187	0.157	(16)[b]	(3)	(24)	(21)
	Adult participants (millions)		Percent increase (decrease) from 2008			
A1B	18.2	20.7	14	31	3	6
	days per participant		Percent increase (decrease) from 2008			
A1B	6.82	5.46	(20)	(8)	(25)	(27)
	total days (millions)		Percent increase (decrease) from 2008			
A1B	124.0	114.2	(8)	21	(23)	(22)

[a] Result of average across CGCM3.1, CSIROMK3.5, and MIROC3.2.
[b] Parentheses denotes decrease.

Floating

The adult participation rate for this non-motorized boating activity—including canoeing, kayaking, tubing, sailing, and whitewater (or other) rafting—averaged about 18.7 percent, or about 18.2 million participants, annually in the North from 2005 to 2009. Across the climate scenarios associated with A1B, the participation rate is expected to decrease 3 to 24 percent by 2060, with the warmer drier scenarios differing from CGCM3.1 by more than 20-percent (Table 22). On average, the 16 percent decrease in participation rate, coupled with population growth, would mean an increase in adult participants of about 14 percent, or 2.5 million people.

Annual days per participant, about 7 in 2008, are expected to decrease across all climate scenarios by 8 to 27 percent, with the drier and warmer climate forecasts roughly tripling the decrease in CGCM3.1. On average, the decrease would be about 20 percent or 1.4 days per participant annually in 2060. Total days of participation will likely increase by 21 percent under CGCM3.1, but decrease by 23 percent under CSIROMK3.5 and 22 percent under MIROC3.2. Thus, depending on the climate changes, annual total days of participation, which totaled 124 million in 2008, could be as low as 95 million or as high as 150 million, although on average a decrease of about 10 million participant days per year is expected.

Key Findings

Outdoor recreation will remain important in the North over the next five decades. The number of participants in 14 of the 17 outdoor recreation activities, or activity aggregates, examined for this report is projected to increase (Table 23). For a number of activities, the per capita participation rate is expected to decrease, but, expected population growth under the A1B simulation (high economic growth and moderate population growth) would be large enough to ensure that only a few—hunting, snowmobiling, and undeveloped skiing—would actually experience a decrease in participants over the next five decades. Of these, snowmobiling and undeveloped skiing could experience large decreases relative to current participant numbers.

In general, participation intensity, or total days of participation, will likely mirror number of participants. Twelve of 17 activities are expected to experience an increase in annual participation days in 2060 compared to 2008 (Table 24). Under A1B, and averaging across the three climate scenarios, non-motorized boating, visiting primitive areas, hunting, snowmobiling, and undeveloped skiing are all likely to experience decreases in total days of participation. Of these, the two winter activities would see the biggest proportional drops, but hunting and visiting primitive areas would experience much larger absolute decreases. More specific discussions of participant numbers, days of participation, and the factors responsible follow.

Table 23. Changes in total outdoor recreation participation across 17 activities by adults in the Northern United States, 2008 to 2060, under Resources Planning Act (RPA) scenario A1B and related climate futures

Activity	2008 Participants	2060 Participant range[a]	2060 Participant range[a]	2060 average Participant change[b]
	(millions)	(millions)	(percent)	(millions)
Developed site use				
Visiting developed sites (family gathering, picnicking, developed camping)	80.5	102 - 109	27 - 35	24
Visiting interpretive sites (nature centers, prehistoric sites, historic sites, other)	67.0	98 - 99	47 - 48	32

Outlook for Outdoor Recreation in the Northern United States 109

Activity	2008 Participants	2060 Participant range[a]	2060 Participant range[a]	2060 average Participant change[b]
	(millions)	(millions)	(percent)	(millions)
Observing nature				
Birding (viewing or photographing)	37.2	46 - 52	24 - 40	11
Nature viewing[c] (viewing or photographing birds, other wildlife, natural scenery, gathering, other)	79.5	105 - 111	32 - 39	28
Backcountry activities				
Challenge (mountain climbing, rock climbing, caving)	9.4	11 - 12	16 - 27	2
Equestrian (horseback riding on trails)	5.8	10 - 12	67 - 108	5
Day hiking	32.4	41 - 44	28 - 35	10
Primitive area use (visiting wilderness, primitive camping, backpacking)	36.1	32 - 44	(11)[d] - 22	1
Motorized activities				
Off-road driving	17.3	20 - 22	18 - 29	7
Motorized snow (snowmobiling)	7.0	2.1 - 4	(70) - (44)	(4)
Motorized water (motor boating, water skiing, personal watercraft use)	26.1	42 - 51	61 - 96	21
Hunting and fishing				
Hunting (all types of legal hunting)	11.3	8 - 11	(28) - 0	(2)
Fishing (warm water, cold water, saltwater, anadromous)	28.7	40 - 49	39 - 71	17
Non-motorized Winter				
Developed skiing (downhill skiing, snowboarding)	11.6	19 - 21	69 - 78	9

J. M. Bowker and Ashley E. Askew

Table 23. (Continued)

Activity	2008 Participants	2060 Participant range[a]	2060 Participant range[a]	2060 average Participant change[b]
	(millions)	(millions)	(percent)	(millions)
Undeveloped skiing (cross-country skiing, snowshoeing)	4.8	3.8 - 4.7	(21) - (2)	(0.6)
Non-motorized Water				
Swimming (screener for various swimming and related activities)	61.7	80 - 91	29 - 47	22
Floating (canoeing, kayaking, rafting, sailing)	18.2	19 - 24	3 - 31	3

[a] Participant range for RPA A1B and climate alternatives (CGCM3.1, CSIROMK3.5, MIROC3.2).
[b] Result of average across CGCM3.1, CSIROMK3.5, and MIROC3.2.
[c] Including birding.
[d] Parentheses denotes decrease or negative value.

Table 24. Changes in total outdoor recreation days across 17 activities by adults in the Northern United States, 2008 to 2060, under Resources Planning Act (RPA) scenario A1B and related climate futures

Activity	2008 days	2060 days range[a]	2060 days range[a]	2060 average days change[b]
	(millions)	(millions)	(percent)	(millions)
Developed site use				
Visiting developed sites (family gathering, picnicking, developed camping)	943	1,054 - 1,201	12 - 27	163
Visiting interpretive sites (nature centers, prehistoric sites, historic sites, other)	516	820 - 887	59 - 72	343
Observing nature				
Birding (viewing or photographing)	3,696	4,413 - 5,006	19 - 35	929
Nature viewing[c] (viewing or photographing birds, other wildlife, natural scenery, gathering, other)	13,925	16,548 - 17,730	19 - 27	3,104

Outlook for Outdoor Recreation in the Northern United States 111

Activity	2008 days	2060 days range[a]	2060 days range[a]	2060 average days change[b]
	(millions)	(millions)	(percent)	(millions)
Backcountry activities				
Challenge (mountain climbing, rock climbing, caving)	37.7	42 - 48	12 - 27	7
Equestrian (horseback riding on trails)	72.3	119 - 144	65 - 100	61
Day hiking	723.8	846 - 923	17 - 28	162
Primitive area use (visiting wilderness, primitive camping, backpacking)	415	282 - 435	(32)[d] - 5	(73)
Motorized activities				
Off-road driving	282.8	291 - 327	3 - 16	24
Motorized snow (snowmobiling)	54.8	15 - 26	(74) - (52)	(35)
Motorized water (motor boating, water skiing, personal watercraft use)	378.8	651 - 827	72 - 118	374
Hunting and fishing				
Hunting (all types of legal hunting)	209.6	121 - 184	(42) - (12)	(64)
Fishing (warm water, cold water, saltwater, anadromous)	515.7	719 - 1,052	39 – 104	404
Non-motorized Winter				
Developed skiing (downhill skiing, snowboarding)	81.3	86 - 113	6 - 39	15
Undeveloped skiing (cross-country skiing, snowshoeing)	32.1	23 - 29	(27) - (10)	(6)
Non-motorized Water				
Swimming (screener for various swimming and related activities)	1,376.2	1,614 - 2,072	17 - 51	396
Floating (canoeing, kayaking, rafting, sailing)	124	96 - 150	(23) - 21	(10)

[a] Days range for RPA A1B and climate alternatives (CGCM3.1, CSIROMK3.5, MIROC3.2).
[b] Result of average across CGCM3.1, CSIROMK3.5, and MIROC3.2.
[c] Including birding.
[d] Parentheses denotes decrease or negative value.

Per Capita Participation

In the next 50 years, under A1B and related climate scenarios for the North, the outdoor recreation activities projected for the most growth in per capita participation (Fig. 7) are developed skiing (25 to 32 percent), horseback riding (23 to 54 percent), fishing (3 to 27 percent), motorized water use (20 to 45 percent), and visiting interpretive areas (9 percent).

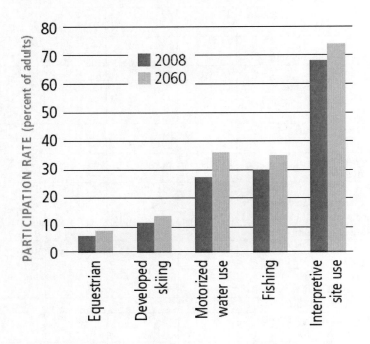

Figure 7. Recreation activities with the highest projected growth in participation rate, 2008 to 2060, in the Northern United States under a future of high economic growth and moderate population growth (A1B) and an average of climate scenarios predicted by three general circulation models: CGCM3.1, CSIROMK3.5, and MIROC3.2. (source: 2010 Resources Planning Act assessment).

A number of activities are projected to experience decreases in adult participation rates. The five activities with the biggest participation rate decreases (Fig. 8) are floating (3 to 24 percent), hunting (26 to 47 percent), snowmobiling (58 to 78 percent), primitive-area visiting (10 to 34 percent), and undeveloped skiing (27 to 41 percent).

Change in participation rates for the remaining activities studied in this report will likely be marginal, vacillating around zero. Generally, activities with currently low per capita rates of participation, such as downhill skiing and horseback riding, have considerable room for growth (decline), but activities with already high participation rates, such as developed site use, nature viewing, and swimming, have less room for growth (decline). Thus, the larger percentage changes in predicted participation rates are often for the currently less popular activities.

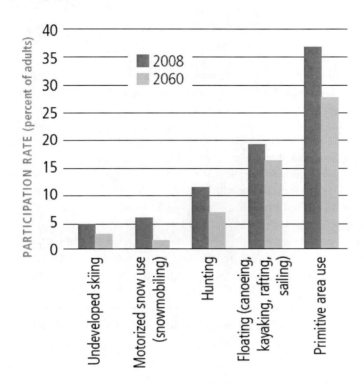

Figure 8. Recreation activities with the lowest projected growth in participation rate, 2008 to 2060, in the Northern United States under a future of high economic growth and moderate population growth (A1B) and an average of climate scenarios predicted by three general circulation models: CGCM3.1, CSIROMK3.5, and MIROC3.2. (source: 2010 Resources Planning Act assessment).

Participant Numbers

Participant numbers follow the predicted trends of participation rates as they are simply the product of participation rate and population. Across climate scenarios (Table 23), the highest growth rates for participant numbers are expected for developed skiing (69 to 78 percent), horseback riding (67 to 108 percent), fishing (39 to 71 percent), motorized water use (61 to 96 percent), and visiting interpretive areas (47 to 48 percent).

A number of activities will likely show less growth and experience decreases in adult participation rates. The five activities that are expected to experience the least growth in participant numbers are non-motorized boating (3 to 31 percent increase), hunting (0 percent growth to 28 percent decrease), snowmobiling (44 to 70 percent decrease), primitive-area visiting (22 percent growth to 11 percent decrease), and undeveloped skiing (2 to 21 percent decrease).

Although growth rates for participant numbers are important, a potentially more important measure for natural resource managers is change in absolute numbers of participants. Activities with already high participation rates often do not demonstrate large percentage changes in participant numbers. However, smaller percentage changes in already highly popular activities can mean quite large changes in the absolute number of adult participants.

The activities that are expected to show the biggest average increases from 2008 to 2060 in participants (Table 23) are visiting developed sites (24 million), nature viewing (28 million), interpretive-area visiting (32 million), swimming (22 million), motorized water use (21 million) and fishing (17 million). These are among the most popular activities examined in this report. Activities expected to have the smallest participant increases across climate scenarios on average, some with participant number decreases, include challenge activities (2 million increase), floating (3 million increase), hunting (2 million decrease), primitive area use (1 million increase), snowmobiling (4 million decrease), and undeveloped skiing (slightly more than 0.5 million decrease).

Participant Days per Year

As described in eq. 3, average activity days per year per participant are used in conjunction with participation rate and population to determine total activity days per year. Yearly days per participant are projected to decrease for most outdoor recreation activities from 2008 to 2060. Three activities, visiting interpretive sites, motorized boating, and fishing are expected to experience increases across the climate scenarios, with average annual days per participant climbing to between 8 and 9 days for visiting interpretive sites and climbing to around 16 days for motorized boating, and to 20 days on average for fishing. Challenge activities will likely maintain about the same number of annual days per participant in 2060 as in 2008.

All other activities are expected to experience a decrease in days per participant per year, with the largest decreases in developed skiing (32 percent), visiting primitive areas (21 percent), and floating (20 percent). For

Outlook for Outdoor Recreation in the Northern United States   115

nature viewing, with a 2008 average of about 176 days per year, a 9 percent decrease by 2060 could translate into an average of 16 fewer activity days per year. However, for activities where participants engage less often, the decreases would be less, less than 1 day per year for snowmobiling and approximately 2 days per year for hunting. For the remaining activities, the changes, although negative, are expected to be relatively minor.

Total Activity Days per Year

Total days are the product of population, participation rate, and days per participant. The five fastest growing outdoor activities, in total days from 2008 to 2060 (Table 24 and Fig. 9), are predicted to be horseback riding (65 to 100 percent), fishing (39 to 104 percent), interpretive-area visiting (59 to 72 percent), motorized water use (72 to 118 percent), and swimming (17 to 51 percent). Alternatively, the five slowest growing activities (Fig. 10) are predicted to be off-road driving (3 to 16 percent increase), primitive area use (5 percent increase to 32 percent decrease), undeveloped skiing (10 to 27 percent decrease), hunting (12 to 42 percent decrease), and snowmobiling (52 to 74 percent decrease).

Higher growth rates do not necessarily imply larger absolute growth. Activities that are currently popular may have slower rates of growth in total days than less popular alternatives, yet their increase in total days may greatly exceed those for less popular but faster growing activities. Averaged over all climate scenarios for A1B, the five activities for which total days would increase the most over the next 50 years (Table 24) are nature viewing (3,104 million days), birding (929 million days), fishing (404 million days), swimming (396 million days) motorized water use (374 million days), and visiting interpretive sites (343 million). Day hiking (162 million days) and visiting developed sites (163 million days) are the only other activities for which days per year are expected to increase by more than 100 million per year by 2060.

Alternatively, five activities are projected to decrease in total activity days per year by 2060 when averaged across all climate scenarios for A1B (Table 24): floating (10 million days), hunting (64 million days), primitive area use (73 million days), snowmobiling (35 million days), and undeveloped skiing (6 million days). These activities are typically space intensive and generally require investments in equipment and training. Moreover, the two winter activities require some level of snow cover.

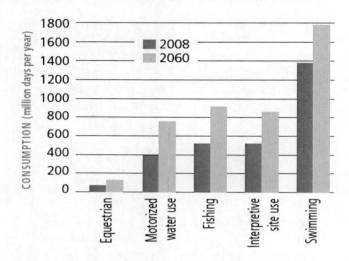

Figure 9. Recreation activities with the highest projected growth in total consumption, 2008 to 2060, in the Northern United States under a future of high economic growth and moderate population growth (A1B) and an average of climate scenarios predicted by three general circulation models: CGCM3.1, CSIROMK3.5, and MIROC3.2. (source: 2010 Resources Planning Act assessment).

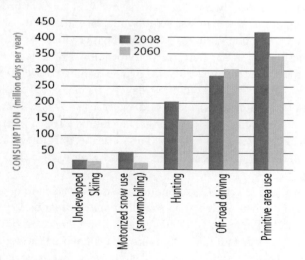

Figure 10. Recreation activities with the lowest projected growth in total consumption, 2008 to 2060, in the Northern United States under a future of high economic growth and moderate population growth (A1B) and an average of climate scenarios predicted by three general circulation models: CGCM3.1, CSIROMK3.5, and MIROC3.2. (source: 2010 Resources Planning Act assessment).

Climate Scenarios

Participant numbers and days of participation were projected for A1B with associated climate scenarios (Fig. 6). Details about climate effects on recreation participation and use can be observed in Tables 6 to 22. No specific probabilities were assigned to any of the three climate scenarios associated with A1B (Joyce et al., in press).

However, the general effects of climate change on each of the 17 outdoor recreation activities examined in this report can be inferred by comparing the percent increases or decreases in Tables 6 to 22. For most activities, the scenarios that are warmer and drier on average over the next five decades (CISROMK3.5 and MIROC3.2) predict lower participation rates, with the largest effects predicted for hunting, snowmobiling, and undeveloped skiing. Alternatively, activities such as horseback riding, motorized water use, and fishing would likely experience relatively higher participation rates under the warmer and drier climate scenarios. For other activities—like visiting interpretive sites, nature viewing, and developed skiing—differences across climate scenarios would be marginal. The general effects of climate change on projections of total days can be similarly observed in the percent increases or decreases in Tables 6 to 22. The pattern is generally the same as with participation.

Other Factors

Examination of model results and odds ratio estimates reveals findings similar to previous research into outdoor recreation participation behavior. First, men are more apt than women to participate in backcountry activities, hunting, fishing, motorized activities, snowmobiling, and floating; and women are more likely to participate in nature viewing, swimming, horseback riding, and visiting interpretive sites.

Ethnicity appears highly associated with participation but it is less a factor on the annual days of participation once an individual has chosen to participate. Minorities, including African Americans, Hispanic Americans, and Asian Americans, are almost always less likely than Caucasians to participate in the activities examined in this report. Respondents claiming Native American, Non-Hispanic identity are often more likely than Caucasians to participate in the remote activities, such as hunting, fishing, off-road driving, snowmobiling, hiking, horseback riding, and nature viewing. These results are similar to previous studies relating ethnicity to recreation participation. A notable exception is day hiking; controlling for other socioeconomic and supply factors, participation is more likely for Hispanic Americans than Caucasians.

Education beyond high school generally resulted in higher participation rates for most activities, but not for all. For example, the higher the education level, the higher the likelihood of participation in birding, non-motorized winter activities, backcountry activities, and nature viewing. However, for fishing and hunting, off-road driving, and snowmobiling, having an education beyond high school is associated with a lowered probability of participation.

Income is positively associated with participation and use across all activities. For some activities— such as birding, hiking, and hunting—the effect was small, but was large for developed skiing and motorized water use. An important aspect of income growth that was not addressed in our analysis for this report should also be mentioned. The RPA assessment variables used in this study were limited to aggregate income growth without regard to changing income distribution over the simulation period. This omission is potentially serious and may overlook the possibility of outdoor recreation access becoming more partitioned by income class.

Relevant land and water availability per capita generally relate positively to activity participation. Hence, decreases in overall forest and rangeland per capita, Federal land per capita, and National Wilderness Preservation System lands per capita produce decreases in spatially intensive activities, such as horseback riding, hunting, motorized off-road driving, visiting primitive areas, and nature viewing.

Similarly, participation in water-based activities, such as swimming, motorized water use, and non-motorized water use, is positively related to the per capita availability of water area. Fishing participation is positively influenced by both water area and forest and rangeland availability. A seemingly counterintuitive result occurs with the variable indicating whether the respondent lives in a coastal community: participation in fishing, hunting, and nature viewing are negatively related to residence in a coastal county, a result that might be driven by the urban dominance of the northern coastal population.

Limitations in Model Results

The model results and projections in this chapter do not account for factors outside the range of available data such as new technology, changes in relative costs, changing site congestion conditions, new infrastructure, acculturation, generational trends, and changes in tastes and preferences. Hence, for an activity like developed skiing, projections of relatively large participant increases could be dampened significantly by declining quality or potentially increased access price that would result from overcrowding combined with declining carrying capacity on areas experiencing climate-induced spatial and temporal limitations. Moreover, the effects of climate on fish and game species were not developed fully enough to include as feedbacks into our behavioral models.

Some other caveats:

- Despite having up to a decade of data for model development, our dataset was not large enough for establishment of any meaningful or statistically significant time-varying parametric relationships. Thus, the participation and days models were static, a substantial limitation when projecting demand over such long time intervals.
- Simulated projections were limited by the quality of the projected exogenous variables.
- The sample of respondents was limited to adults (16 years and older); thus, the effects of recreation demand by youth were omitted. For activities that are traditionally adult in nature—such as challenge sports, visiting primitive areas, and hunting— omitting children is likely not a serious omission. However, for visiting developed sites, swimming, fishing, visiting interpretive areas, and other family-oriented activities, the results presented herein could be biased somewhat downward relative to overall use.
- By performing the analyses at the regional level, we may have overlooked important subregional changes and resulting implications. For example, visiting a primitive or wilderness area or day hiking may have a different meaning for a rural resident than an urbanite.

CONCLUSION

Under the demographic, land use, and climate conditions that we considered for this report, recreation participant numbers and days in the field will likely grow for most activities over the next 50 years. Thus, the general outlook for outdoor recreation resources in the North is a per capita reduction in opportunities and access.

Assuming that the public land base for outdoor recreation remains stable and the privately owned land base available for recreation decreases, an increasing population would result in decreasing opportunities for recreation per person across most of the region. Although many other factors are involved in recreation supply, recreation resources (both natural and constructed) likely will become less "available" as more people compete to use them.

On privately owned land, increased competition for recreational resources resulting from increased demand relative to supply could mean rising access prices. On public lands, where access fees cannot be adjusted easily to market or quasi-market conditions, increased congestion and possible decreases in the quality of the outdoor recreation experience are likely to present important challenges to management.

A major challenge for natural resource managers and planners will be to ensure that recreation opportunities remain viable and that they grow along with the population. This challenge will probably have to be met through creative and efficient management of site attribute inputs and plans, rather than through any major expansions or additions to the natural resource base. Trends toward more flexible work scheduling and telecommuting may allow recreationists to allocate their leisure time more evenly across the seasons and through the week, thus facilitating less concentrated peak demands. As well, entrepreneurs and managers may identify opportunities for more efficiently using resources by expanding seasonal opportunities, for example developing mountain biking venues at ski areas.[†] Conversely, such technological innovations as global positioning system units and inexpensive plastic kayaks would allow people to find and get to places more easily and quickly, perhaps leading to overuse pressures not previously considered a threat.

Overall, a future in which the infrastructure supporting the region's outdoor recreation opportunities will not be severely tested is hard to envision. For activities like improved-facility use and day hiking, fewer acres or trail miles per participant could begin to strain existing infrastructure as biological and social carrying capacities are exceeded. Some activities may not require expansive contiguous areas for quality experiences; examples are birding, which is often "edge dependent," and hiking, which occurs along linear corridors. However, activities typically considered space intensive— horseback riding on trails, motorized water use, fishing, and off-road

---

[†] D.J. Mansius, director, Forest Policy & Management, Maine Forest Service, Dept. of Conservation. Personal communication. August 17, 2012.

driving—are likely to actually "feel" much more congested given the nature of the activity. Alternatively, activities like hunting and primitive area use could feel considerably less congested because of projected decreases in annual user days.

Measures of use per acre or other units of infrastructure are not comparable across recreation activities, and some may actually have a social component—with more congestion yielding increased user utility—but only up to a point. For activities that may be near carrying capacity from a recreation user perspective, or infrastructure carrying capacity, large increases in use per acre could be a concern, both for the land and for the user. Increased pressure can be expected on fishery resources, water quantity, and water quality in areas used for motor boating and at interpretive areas.

Because general forest area recreation usage— including hunting, off-road driving, fishing, and horseback riding on trails—generally require more space per user for high quality (and safe) experiences, an increase in use density would undoubtedly be of concern to national forest managers. For example, conflicts arising from congestion may increase, not only within an activity (such as off-road drivers running into each other figuratively and literally) but also across activities (such as off-road drivers spooking horses and scaring away game sought by hunters). Managers of general forest areas may have to choose among potentially unpopular access regulation schemes to mitigate conflicts. They may also need to consider sectioning general forest areas into special use areas for specific activities—such as off-road driving, horseback riding on trails, and hunting—to reduce cross-activity congestion conflicts. Needless to say, the increased congestion can only increase the impacts of recreation on the forest environment.

Choices in outdoor recreation activities have changed over time in response to changing tastes and preferences, demographics, technological innovations, economic conditions, and changing recreational opportunities. Overall, the number of nature-based outdoor recreation participants has increased since the last RPA assessment, continuing a long-term trend. At the same time, recreation visitation to State parks and Federal lands has apparently not increased at similar rates, indicating that recreationists are also using other resources.

The change in recreation preferences at least partly reflects changing demographics in the U.S. public. As the northern population ages and becomes more racially and ethnically diverse, no one can predict with certainty how future recreation demand and supply will adjust. Based on the available data, we nevertheless project future growth for most recreation activities. Future demand, of course, can be expected to change as relative costs, competition for access, and other scarcity factors change and affect choices for recreation activities, times, and locations.

Climate can affect willingness to participate in recreation activities as well as recreation resource availability and quality. The climate variables that we used in the recreation models were limited to those from the RPA assessment or were derivatives of those basic variables. Generally, these variables were presumed to affect willingness to participate and frequency of participation directly. However, even without existing data, climate change might be expected to affect resource availability, directly and indirectly. For example, increasing temperatures will likely affect the distribution of plant and animal species fundamental to maintaining fish and game populations. Moreover, changes in precipitation may influence local snow cover and thus affect seasonal availability for such activities as snowmobiling and undeveloped skiing. Walls et al. (2009) concluded that the single most important new challenge to recreation supply will be mitigating the adverse effects of climate change, particularly in coastal areas and on western public lands. Because disentangling the effects of the climate variables on recreation participation is difficult, further exploration of these direct and indirect relationships—at both local and macro levels—will be fundamental to improving future forecasts.

REFERENCES

Alcamo, J.; Ash, N.J.; Butler, C.D.; Let al.]. 2003. Ecosystems and human well-being. Washington, DC: Island Press. 245 p.

Bowker, J.M. 2001. Outdoor recreation participation and use by Alaskans: projections 2000-2020. Gen. Tech. Rep. PNW–GTR–527. Portland, OR: U.S. Department of Agriculture, Forest Service, Pacific Northwest Research Station. 28 p.

Bowker, J.M.; Askew, A.E.; Cordell, H.K.; Bergstrom, J.C. In press. Outdoor Recreation. In: Wear, D.N. and J.G. Greis, eds. Southern Forest Futures Project. Gen. Tech. Rep. Asheville, NC: U.S. Department of Agriculture, Forest Service, Southern Research Station.

Bowker, J.M.; Askew, A.E.; Cordell, H.K.; Let al.]. 2012. Outdoor Recreation Participation in the United States – Projections to 2060: A technical document supporting the Forest Service 2010 Resources Planning Act Assessment. Gen. Tech. Rep. SRS-160. Asheville, NC: U.S. Department of Agriculture, Forest Service, Southern Research Station. 36 p.

Bowker, J.M.; Bergstrom, J.C.; Gill, J. 2007. Estimating the economic value and impacts of recreational trails: a case study of the Virginia Creeper rail trail. Tourism Economics. 13: 241–260.

Bowker, J.M.; Bergstrom, J.C.; Starbuck, C.M. Let al.]. 2010. Estimating demographic and population level induced changes in recreation demand for outdoor recreation on US national forests: an application of National Visitor Use Monitoring Program data. Fac. Ser. Work. Pap. FS 1001. Athens, GA: University of Georgia, Department of Agricultural and Applied Economics. 147 p.

Bowker, J.M.; English, D.B.K.; Cordell, H.K. 1999. Outdoor recreation participation and consumption: projections 2000 to 2050. In: Cordell, H.K.; Betz, C.J.; Bowker, J.M.; [et al.]. Outdoor recreation in American life: a national assessment of demand and supply trends. Champagne, IL: Sagamore Press: 323–350.

Bowker, J.M.; Leeworthy, V.R. 1998. Accounting for ethnicity in recreation demand: a flexible count data approach. Journal of Leisure Research. 30: 64 –78.

Bowker, J.M.; Murphy, D.; Cordell, H.K.; [et al.]. 2006. Wilderness and primitive area recreation participation and consumption: an examination of demographic and spatial factors. Journal of Agricultural and Applied Economics. 38(2): 317–326.

Cameron, C.A.; Trivedi, P.K. 1998. Econometric society monographs: regression analysis of count data. New York: Cambridge University Press. 412 p.

Cicchetti, C.J. 1973. Forecasting recreation in the United States. Lexington, MA: D.C. Heath and Co. 200 p.

Cordell, H.K., ed. 2012. Outdoor recreation trends and futures: a technical document supporting the Forest Service 2010 RPA assessment. Gen. Tech. Rep. SRS–150. Asheville, NC: U.S. Department of Agriculture, Forest Service, Southern Research Station. 167 p.

Cordell, H.K.; Bergstrom, J.C. 1991. A methodology for assessing national outdoor recreation demand and supply trends. Leisure Sciences. 13(1): 1–20.

Cordell, H.K.; Bergstrom, J.C.; Hartmann, L.A.; English, D.B.K. 1990. An analysis of the outdoor recreation and wilderness situation in the United States: 1989-2040. Gen. Tech. Rep. RM–189. Fort Collins, CO: U.S. Department of Agriculture, Forest Service, Rocky Mountain Forest and Range Experiment Station. 112 p.

Cordell, H.K.; Betz, C.J.; Zarnoch, S.J. 2013. Recreation and protected land resources in the United States: a technical document supporting the Forest Service 2010 RPA Assessment. Gen. Tech. Rep. SRS–169. Asheville, NC: U.S. Department of Agriculture, Forest Service, Southern Research Station. 198 p.

Cordell, H.K.; Betz, C.J.; Mou, S.H.; Gormanson, D. 2012. Outdoor recreation in a shifting northern societal landscape. Gen. Tech. Rep. NRS-100. Newtown Square, PA: U.S. Department of Agriculture, Forest Service, Northern Research Station. 74 p.

Dale, D.; Weaver, T. 1974. Trampling effects on vegetation of the trail corridors of north Rocky Mountain forests. Journal of Applied Ecology. 11: 767-772.

Englin, J.E.; Shonkwiler, J.S. 1995. Estimating social welfare using count data models: an application to long-run recreation demand under conditions of endogenous stratification and truncation. Review of Economics and Statistics. 77(1): 104–112.

English, D.B.K.; Betz, C.J.; Young, M.J. 1993. Regional demand and supply projections for outdoor recreation. Gen. Tech. Rep. RM–230. Fort Collins, CO: U.S. Department of Agriculture, Forest Service, Rocky Mountain Forest and Range Experiment Station. 44 p.

English, D.B.K.; Kocis, S.M.; Zarnoch, S.J.; Arnold, J.R. 2002. Forest Service national visitor use monitoring process: research method documentation. Gen. Tech. Rep. SRS–57. Asheville, NC: U.S. Department of Agriculture, Forest Service, Southern Research Station. 14 p.

Greene, W.H. 2009. NLOGIT 4.0. Plainview, NY: Econometric Software, Inc.

Hagerty, D.; Moeltner, K. 2005. Specification of driving costs in models of recreation demand. Land Economics. 81(1):127-143.

Hall, C.M.; Page, S.J. 1999. The geography of tourism and recreation. New York, NY: Routledge. 309 p.

Hall, T.E.; Heaton, H.; Kruger, L.E. 2009. Outdoor recreation in the Pacific Northwest and Alaska: trends in activity participation. Gen. Tech. Rep. PNW–778. Portland, OR: U.S. Department of Agriculture, Forest Service, Pacific Northwest Research Station. 108 p.

Hof, J.G.; Kaiser, H.F. 1983a. Long term outdoor recreation participation projections for public land management agencies. Journal of Leisure Research. 15(1): 1–14.

Hof, J.G; Kaiser, H.F. 1983b. Projections of future forest recreation use. Resour. Bull. WO–2. Washington, DC: U.S. Department of Agriculture, Forest Service. 12 p.

Intergovernmental Panel on Climate Change (IPCC). 2007. Climate change 2007, the fourth IPCC assessment report. http://www.ipcc.ch /publications_and_data/publications_and_data_ reports.shtml (10 January 2013).

Joyce, L.A.; Price; D.T.; Coulson, D.P.; [et al.]. [In press]. Projecting climate change in the United States: a technical document supporting the Forest Service 2010 RPA Assessment. Gen. Tech. Rep. RMRS. Fort Collins, CO: U.S. Department of Agriculture, Forest Service, Rocky Mountain Research Station.

Leeworthy, V.R.; Bowker, J.M.; Hospital, J.D.; Stone, E.A. 2005. Projected participation in marine recreation: 2005 & 2010. Report prepared for U.S. Department of Commerce, National Oceanic and Atmospheric Administration, National Ocean Service, Special Projects Division, Silver Spring, MD. 152 p. Available at http://www.srs.fs.usda.gov/pubs/ja /ja_leeworthy002.pdf (10 January 2013).

Manning, R. E. 1997. Social carrying capacity of parks and outdoor recreation areas. Parks and Recreation. 32(10): 32-38.

Nakic'enovic', N.; Alcamo, J.; Davis, G.; Let al.]. 2000. Emissions scenarios. a special report of working group III of the Intergovernmental Panel on Climate Change. Cambridge, United Kingdom and New York, NY, USA: Cambridge University Press. 599 p. Available at http://www.grida.no /climate/ipcc/emission/index.htm (10 January 2013).

Ovaskainen, V.; Mikkola, J.; Pouta, E. 2001. Estimating recreation demand with on-site data: an application of truncated and endogenously stratified count data models. Journal of Forest Economics. 7(2): 125–144.

Poudyal, N.C.; Cho, S.H.; Bowker, J.M. 2008. Demand for resident hunting in the southeastern United States. Human Dimensions of Wildlife 13: 154-178.

SAS Institute Inc. 2004. SAS/STAT 9.1 user's guide. Cary, NC: SAS Institute Inc. 5,121 p.

Souter, R.A.; Bowker, J.M. 1996. A note on nonlinearity bias and dichotomous choice CVM: implications for aggregate benefits estimation. Agricultural and Resource Economics Review. 25(1): 54–59.

U.S. Census Bureau. 2004. U.S. interim projections by age, sex, race and Hispanic origin. http://www.census.gov/population/projections/. Washington, DC: U.S. Department of Commerce, Census Bureau. (10 January 2013).

U.S. Department of Agriculture Forest Service. 2009. National survey on recreation and the environment [Dataset]. Asheville, NC: U.S. Department of Agriculture, Forest Service. Available at www.srs.fs.usda.gov/trends /nsre/nsre2.html. (15 September 2010).

U.S. Department of Agriculture Forest Service. 2010. National Visitor Use Monitoring Program: FY 2009 NVUM national summary report. Washington, DC: U.S. Department of Agriculture, Forest Service. Available at www.fs.fed.us/recreation /programs/nvum/. (6 October 2010).

U.S. Department of Agriculture Forest Service. 2012. Future Scenarios: A technical document supporting the Forest Service 2010 RPA Assessment. Gen. Tech. Rept. RMRS–272. Fort Collins, CO: U.S. Department of Agriculture, Forest Service, Rocky Mountain Research Station. 34 p.

U.S. Department of Commerce Bureau of Economic Analysis. 2008a. National and income product accounts table 1.15. Gross domestic product. Version January 30, 2008. Washington, DC: U.S. Department of Commerce, Bureau of Economic Analysis.

U.S. Department of Commerce Bureau of Economic Analysis. 2008b. National and income product accounts table 2.1. Personal income and its disposition. Version January 30, 2008. Washington, DC: U.S. Department of Commerce, Bureau of Economic Analysis.

Walls, M.; Darley, S.; Siikamaki, J. 2009. The state of the great outdoors: America's parks, public lands, and recreation resources. Washington, DC: Resources for the Future. 97 p.

Walsh, R.G.; Jon, K.H.; McKean, J.R.; Hof, J. 1992. Effect of price on forecasts of participation in fish and wildlife recreation: an aggregate demand model. Journal of Leisure Research. 21: 140–156.

Wear, David N. 2011. Forecasts of county-level land uses under three future scenarios: a technical document supporting the Forest Service 2010 RPA Assessment. Gen. Tech. Rep. SRS–141. Asheville, NC: U.S. Department of Agriculture, Forest Service, Southern Research Station. 41 p.

Zarnoch, S.J.; Cordell, H.K.; Betz, C.J.; langner, l. 2010. Projecting county-level populations under three future scenarios: a technical document supporting the Forest Service 2010 RPA assessment. e-Gen. Tech. Rep. SRS–128. Asheville, NC: U.S. Department of Agriculture, Forest Service, Southern Research Station. 8 p.

Zawacki, W.T.; Marsinko, A.; Bowker, J.M. 2000. A travel cost analysis of economic use value of nonconsumptive wildlife recreation in the United States. Forest Science. 46(4): 496–505.

INDEX

#

21st century, 55, 57

A

access, 32, 39, 45, 46, 53, 55, 56, 57, 118, 119, 120, 121, 122
accounting, 81
acculturation, 119
activity level, 19
adults, 55, 62, 84, 86, 91, 92, 93, 94, 95, 96, 97, 98, 99, 100, 101, 102, 103, 104, 105, 106, 107, 108, 110, 119
adverse effects, 122
African Americans, 5, 117
age, vii, 1, 2, 7, 8, 17, 19, 22, 23, 24, 26, 27, 30, 31, 50, 51, 59, 61, 62, 63, 66, 69, 72, 84, 91, 125
agencies, 2, 32, 33, 34, 40, 43, 48, 53, 63, 64, 69, 124
aggregate demand, 126
aging population, 99
Alaska, 9, 32, 33, 34, 39, 48, 61, 84, 124
Asian Americans, 117
assessment, viii, 2, 3, 62, 67, 68, 73, 74, 75, 76, 77, 78, 79, 82, 83, 112, 113, 116, 118, 121, 122, 123, 125, 126
automobiles, 3

B

base, 40, 59, 65, 76, 84, 120
base year, 40, 65, 76
behavioral models, 119
behaviors, 73
benefits, 56, 125
bias, 125
big game hunting, 4, 20, 24
biodiversity, 2
bioenergy, 73
biomass, 74
birding, 22, 71, 92, 93, 110, 111, 115, 118, 120
birds, vii, 4, 18, 19, 22, 23, 52, 70, 93, 101, 109, 110
Bureau of Land Management, 33, 34, 45, 48, 59, 63, 84
businesses, 65

C

case study, 123
category a, 10, 91
Caucasians, 96, 117
CBP, 65
Census, 5, 6, 7, 9, 10, 11, 12, 14, 15, 16, 17, 23, 24, 26, 27, 33, 34, 37, 40, 41, 42, 43, 44, 47, 49, 57, 58, 59, 60, 61, 64, 65, 66, 72, 74, 81, 83, 125

128 Index

challenges, 120
Chicago, 10, 11, 16, 43, 44, 50
children, 25, 29, 30, 55, 63, 119
circulation, 73, 79, 112, 113, 116
city(s), 9, 10, 44, 46, 51
citizens, 6
classes, 10, 13, 29, 30, 31, 50, 52, 65, 76
classification, 35, 36
climate(s), viii, 2, 55, 67, 68, 72, 73, 79, 80,
    82, 85, 86, 88, 89, 90, 91, 92, 93, 94, 95,
    96, 97, 98, 99, 100, 101, 102, 103, 104,
    105, 106, 107, 108, 110, 111, 112, 113,
    114, 115, 116, 117, 119, 122, 125
climate change, 2, 79, 91, 102, 107, 117,
    122, 125
closure, 41
clusters, 16
commercial, 55
communication, 120
community(s), 55, 118
competition, 122
complement, 69
composites, 69, 71, 94
composition, vii, 1, 2, 5, 13, 60
Congress, 3, 34
conservation, 25, 36, 43, 53
consumption, 71, 72, 80, 82, 83, 84, 86,
    116, 123
convergence, 74
cooperation, 63
cost, 126

D

data collection, 17
database, 56, 58, 63, 64, 83, 86
death rate, 9
demographic data, 60, 65, 66
Department of Agriculture, 1, 34, 58, 66,
    67, 126
Department of Commerce, 24, 26, 27, 58,
    59, 126
Department of the Interior, 33, 34, 48, 59,
    60, 63
dependent variable, 85

derivatives, 122
discontinuity, 76
disposition, 126
distribution, 8, 32, 38, 60, 122
District of Columbia, 13, 14, 51, 61
diversity, 38, 54, 56, 97
dominance, 118

E

economic downturn, 76, 81
economic growth, 73, 74, 76, 77, 78, 79, 88,
    108, 112, 113, 116
ecosystem, 56
education, 37, 40, 43, 64, 101, 104, 118
electricity, 77, 78
emission, 125
employees, 65
employment, 50, 55, 59, 64
employment opportunities, 50, 55
energy, 74
entrepreneurs, 120
environment, 58, 60, 66, 121, 125
equipment, vii, 4, 25, 115
ethnicity, 5, 6, 8, 61, 72, 99, 117, 123
evapotranspiration, 86, 87, 88

F

families, 55
fauna, 70, 94
financial, 64
fish, 4, 22, 23, 24, 37, 40, 52, 64, 102, 119,
    122, 126
Fish and Wildlife Service, 33, 48, 59, 60, 63
fishing, 4, 20, 21, 22, 24, 26, 29, 30, 31, 52,
    70, 101, 102, 103, 109, 111, 112, 113,
    114, 115, 117, 118, 119, 120, 121
flora, 70, 94
flowers, 22, 23, 52
force, 77
forecasting, 84, 86, 89
forest resources, 78

Index

G

GDP, 74, 75, 76
geography, 124
Georgia, 9, 41, 62, 63, 123
global recession, 76
governments, 59
greenhouse, 79
greenhouse gas, 79
greenhouse gas emissions, 79
gross domestic product, 74, 75, 76
group activities, 5
grouping, 28
growth, vii, 1, 3, 5, 8, 10, 11, 12, 13, 14, 15, 16, 17, 22, 24, 35, 39, 40, 43, 50, 51, 52, 53, 55, 62, 74, 75, 77, 81, 86, 88, 91, 95, 96, 98, 101, 104, 105, 108, 111, 112, 113, 114, 115, 116, 118, 121
growth rate, 5, 8, 43, 50, 51, 53, 75, 101, 113, 114, 115

H

Hawaii, 41, 48, 61
high school, 84, 118
highways, 3
hiking trails, 69
Hispanic population, 5, 11, 12, 13, 50, 61
Hispanics, 5, 11, 96
history, 55, 64, 72, 78
homes, 53
horses, 121
household income, 84, 104
housing, 58
human, 122
hunting, viii, 4, 20, 21, 26, 27, 52, 68, 101, 102, 108, 109, 111, 112, 114, 115, 117, 118, 119, 121, 125

I

identity, 117
income, 14, 15, 16, 60, 66, 72, 75, 77, 80, 104, 118, 126

income distribution, 118
increased access, 119
increased competition, 120
Independence, 58
individuals, 19, 50, 55, 82
information exchange, 57
infrastructure, 55, 119, 120, 121
institutions, 44, 84
investments, 25, 115
Iowa, 11, 12, 14, 17, 44, 46, 51, 61
irrigation, 43
issues, 2

J

Jordan, 57
jurisdiction, 39, 40

L

lakes, vii, 4, 22, 23
land use, viii, 2, 58, 67, 68, 72, 76, 77, 80, 85, 86, 88, 119, 126
landscape, 55, 124
lead, viii, 13, 68, 96, 97
learning, 22, 43
leisure, 3, 64, 120
leisure time, 120
light, 72
linear model, 83
local government, 36, 40, 42, 43, 53, 59, 64

M

majority, 38, 44, 72
management, 37, 39, 47, 57, 63, 65, 120, 124
manufacturing, 3
Maryland, vii, 1, 2, 13, 14, 44, 51, 61, 68
media, 29, 30
methodology, 123
metropolitan areas, 9, 10, 12, 17, 46, 54, 55
migration, 57
Minneapolis, 9, 50

130 Index

Mississippi River, 11
Missouri, vii, 1, 2, 3, 12, 13, 14, 16, 17, 45, 50, 54, 61, 62, 68
models, 72, 73, 79, 80, 81, 82, 83, 85, 86, 88, 89, 90, 112, 113, 116, 119, 122, 124, 125
music, 28, 29, 31

N

national character, 74
National Park Service, 33, 34, 45, 48, 60, 63
National Survey, 3, 62, 71, 81
natural resources, 3, 25, 36, 43
New England, 12, 15, 53
next generation, 55
niche market, 25
North America, 65
northern forest, vii, 1, 2, 4, 57
NPS, 48
NRS, 1, 58, 124

O

omission, 118, 119
operations, 64
opportunities, 2, 9, 38, 39, 43, 55, 119, 120, 121
ownership, 32

P

Pacific, 5, 6, 7, 8, 9, 13, 14, 19, 20, 21, 32, 33, 34, 35, 36, 37, 38, 39, 40, 41, 42, 44, 48, 49, 50, 78, 79, 84, 122, 124
Pacific Islanders, 5
parallel, 72
parameter estimates, 83, 85, 88
parents, 55
participants, viii, 18, 19, 21, 22, 23, 24, 26, 27, 51, 52, 56, 67, 69, 71, 73, 80, 82, 88, 90, 91, 92, 93, 94, 95, 96, 97, 98, 99, 100, 101, 102, 103, 104, 105, 106, 107, 108, 114, 115, 121

pasture, 76
payroll, 59, 65
Philadelphia, 50
plants, 4, 19, 22, 23, 52, 92
playing, 29, 31
pleasure, 22, 23, 51
policy, 2
policy makers, 2
ponds, 4, 22, 23
pools, 23, 40, 41
population density, 10, 11, 15, 16, 51, 61, 94, 97, 101
population growth, viii, 5, 10, 12, 13, 14, 15, 16, 17, 32, 47, 48, 51, 54, 55, 62, 67, 68, 72, 73, 74, 77, 78, 79, 80, 81, 88, 90, 91, 93, 95, 96, 97, 98, 99, 101, 102, 103, 104, 107, 108, 112, 113, 116
precipitation, 79, 86, 87, 104, 122
price changes, 81
probability, 72, 81, 82, 85, 88, 89, 118
project, vii, 72, 81, 83, 121
proliferation, 3
protected areas, 37
protection, 34
public interest, 69
public service, 56
public support, 69

R

race, 6, 8, 59, 61, 66, 72, 125
radius, 43, 65
rainfall, 101
rangeland, 76, 78, 79, 84, 118
rate of change, 13, 32
reading, 29, 30
real estate, 63
real income, 3
recession, 25, 39
recreational, 35, 43, 120, 121, 123
regionalism, 74
regression, 123
regression analysis, 123
researchers, 81
resource availability, 65, 73, 122

Index

resource management, 69

resources, vii, 1, 2, 4, 10, 35, 36, 38, 53, 54, 56, 58, 63, 64, 65, 71, 75, 119, 120, 121, 124, 126

response, 79, 121

restrictions, 32

rowing, 4, 20

rural areas, 56

S

salmon, 4

saltwater, 4, 20, 24, 70, 102, 103, 109, 111

SAS, 82, 125

scarcity, 56, 122

scattering, 12

school, 25, 84, 118

science, 32, 37

sensitivity, 62

services, 39, 40, 41, 43, 53, 56, 64

sewage, 43

sex, 59, 66, 125

showing, 60, 61, 98, 105

simulation, 73, 80, 108, 118

social welfare, 124

society, 25, 123

species, 119, 122

specifications, 83

spending, 25, 28, 29, 55

Spring, 56, 66, 125

staffing, 39

state(s), 2, 17, 18, 38, 48, 57, 58, 59, 60, 66, 68, 126

statistics, 59, 61, 72, 88

stratification, 124

stretching, 44, 46, 50

structure, 81, 85

surface area, 48, 49, 54, 65

sustainability, vii, 1, 2

T

team sports, 29, 30, 31

technological change, 81

technology, 25, 118

teens, 30

telephone, 62, 63

temperature, 79, 80, 86, 87, 104

Tennessee Valley Authority, 32, 33, 45, 48, 58, 63

territorial, 47, 49

threats, 2

tourism, 124

trade, 74

training, 115

TVA, 48

U

U.S. Army Corps of Engineers, 32, 33, 45, 48, 58, 63

U.S. Department of Agriculture, vii, 1, 2, 18, 19, 20, 23, 24, 26, 27, 33, 38, 41, 45, 46, 48, 49, 57, 58, 60, 61, 62, 64, 66, 73, 75, 76, 77, 82, 85, 86, 122, 123, 124, 125, 126

U.S. Department of Commerce, 5, 6, 7, 9, 10, 11, 12, 14, 15, 16, 17, 23, 24, 26, 27, 33, 34, 37, 40, 41, 42, 44, 47, 49, 57, 58, 59, 61, 66, 75, 125, 126

U.S. Department of the Interior, 32, 33, 34, 35, 37, 48, 59, 60, 63

U.S. regions, vii, 1, 2, 19

United Kingdom, 125

urban, 2, 9, 10, 11, 15, 16, 35, 44, 45, 50, 56, 76, 77, 118

urban areas, 35

urbanization, 77, 78

USA, 125

V

valuation, 83

variables, 72, 73, 76, 80, 81, 82, 83, 84, 85, 86, 88, 89, 90, 118, 119, 122

variations, 99

vector, 82, 83, 85

vegetation, 124

vehicles, 29, 30, 31, 37

W

walking, 28, 29, 30, 31, 51, 52
Washington, 9, 11, 12, 15, 34, 50, 51, 57, 58, 59, 60, 66, 122, 124, 125, 126
water, vii, 1, 2, 4, 10, 22, 29, 31, 32, 43, 46, 47, 49, 52, 54, 55, 56, 58, 63, 65, 70, 76, 77, 78, 81, 89, 98, 100, 101, 103, 106, 109, 111, 112, 113, 114, 115, 117, 118, 120, 121
water quality, 121
water resources, vii, 1, 32, 54, 56, 63
well-being, 122
wetlands, 43

whitewater activities, viii, 68
wilderness, 3, 4, 19, 23, 34, 53, 56, 60, 70, 85, 97, 98, 109, 111, 119, 123
wildland, 5
wildlife, vii, 2, 4, 18, 19, 22, 23, 29, 30, 31, 32, 37, 40, 51, 52, 64, 101, 109, 110, 126
Wisconsin, 14, 16, 43, 44, 45, 46, 54, 61
World War I, 3

Y

yes/no, 64, 82
yield, 82, 88, 90
young adults, 8
young people, 9, 25, 52, 55